Accession no
755798
LG
01
WITHDRAWN

Chester
A College of the
University of Liverpool

D1643062

The Locomotion of Soft-Bodied Animals

E. R. Trueman

D. Sc., F.I.Biol.
Beyer Professor of Zoology in the University of Manchester

Edward Arnold

First published 1975
by Edward Arnold (Publishers) Limited
25 Hill Street, London W1X 8LL

Boards Edition ISBN: 0 7131 2485 7
Paper Edition ISBN: 0 7131 2486 5

Printed in Great Britain by
J. W. Arrowsmith Ltd.,
Bristol, England

Preface

The ability to move is a fundamental character of animals. Analysis of the locomotory mechanics of arthropods and chordates has been facilitated by the presence of a rigid skeleton whereas studies of soft-bodied animals have often encountered experimental difficulties. In recent years, however, the development of electronic techniques for recording movement and pressure has produced significant advances in our knowledge of the locomotion of the lower metazoans, including the Mollusca. In this book locomotory mechanisms are discussed in relation to habitat and the form of animals, particular attention being drawn to the mechanism of muscular locomotory waves for movement over firm surfaces, the penetration of soft and hard substrates, swimming by undulatory waves and by jet propulsion and the functional importance of different body cavities.

Most major steps in metazoan history are concerned with changes in body organisation and here mechanical factors are of prime importance. Thus a grasp of the problems of locomotion in different habitats is fundamental to the understanding of the origin and later evolution of animal phyla. For this reason, some attention is given to theories of the phylogeny of soft-bodied animals including the origin and evolution of the Metazoa, which should be brought to the attention of all students of animal biology. In writing this book I have assumed that the reader will already have a basic knowledge of invertebrates which may be reinforced by reference to one of the recently published textbooks of invertebrate anatomy and classification, for example the *Textbook of Zoology, Invertebrates,* by A. J. Marshall and W. D. Williams (Macmillan, London).

I owe much and am most grateful to many friends and colleagues who have very willingly discussed and investigated problems with me. They include, Drs A. D. Ansell, A. R. Brand, Professor A. C. Brown, Drs M. E. G. Evans, E. J. Iles, A. Packard and P. D. Soden. It is a pleasure to thank Mr A. Packard for his photographs of *Octopus* and Dr H. D. Jones for his continuing support and for supplying photographs used in illustrating this book; Professor E. J. W. Barrington, F.R.S., the zoological editor of this series, for his advice and encouragement; and my wife for her encouragement and support over many years not only while writing this book but also during my experimental investigations which form the basis of my interest in this field.

I am most grateful to the following for permission to copy illustrations from their publications and to the authors whose names are cited in the legends: Biologische Anstalt Helgoland, Figs. 6.14 and 6.15; Company of Biologists Ltd., Figs. 1.6, 2.9, 2.10, 3.2, 3.3, 3.9, 3.16, 5.14, 6.3, 6.4, 6.6 and 6.7; George Allen and Unwin Ltd., Figs. 3.1 and 3.7; Macmillan (Journals) Ltd., Fig. 1.7; Marine biological Ass. U.K., Fig., 7.7; Marine biological Laboratory, Woods Hole, Fig. 3.11; North Holland Publishing Company, Fig. 4.3; the Royal Society, Figs. 3.16, 4.6, 4.7, 4.8, 4.9 and 4.12; the Zoological Society, London, Figs. 3.6, 3.10, 3.19, 3.20, 4.11, 5.2, 5.4 and 5.5. Finally I am most indebted to Mrs J. Gatehouse and Mrs J. Murphy for so carfully typing the manuscript of this book, often from illegible drafts, and to Mr L. Lockey for photographic assistance.

Manchester, 1974 E. R. T.

Contents

PREFACE v

1 THE MECHANICAL PRINCIPLES 1
Principles of locomotion 2
Hydraulic systems 4
The determination of pressure 8
The skeleton of sea anemones 10
The hydraulic system of ascidians 14

2 MOVEMENT OVER HARD SURFACES 16
Ciliary locomotion in flatworms and nemertines 16
Patterns of muscular activity 18
The form and direction of propagated waves 25
Pedal crawling in snails 27

3 THE PENETRATION OF THE SUBSTRATE BY SOFT-BODIED ANIMALS 42
Introduction 42
Mechanical principles of burrowing 43
Burrowing activity 47
Initial penetration of substrate 48
Movement through a soft substrate 55

Comparative survey of burrowing 70
Fluid dynamics of burrowing 75
Control of burrowing 80
Migratory behaviour of *Donax* 84

4 BORING INTO HARD SUBSTRATES 87
The mechanisms of rock boring 88
Boring activities of bivalves 89
Evolution of the boring habit 104

5 CRAWLING AND UNDULATORY SWIMMING 107
Introduction 107
Crawling 107
Undulatory swimming 114
Undulatory swimming with parapodia 122
Segmentation and swimming 124

6 SWIMMING BY JET PROPULSION 129
Introduction 129
Jet propulsion in the Cephalopoda 130
Mantle muscles during the jet cycle 142
Jet propulsion in other soft-bodied invertebrates 150
Conclusions 157

7 LOCOMOTION AND METAZOAN EVOLUTION 159
Introduction 159
An outline of early metazoan evolution 161
Protostomia and Deuterostomia 176
Origins of the major phyla 179

REFERENCES 188

INDEX 195

I

The Mechanical Principles

An essential requirement for the movement of animals is the means for affording muscular antagonism. Any system which does this may be described as a skeleton, its function being to ensure that muscles act upon each other for restitution of their relaxed state after contraction. From this aspect the locomotion of animals and the mechanical principles involved fall into two broad categories, although the detail may differ even between species. First, and more primitive, are the mechanisms of soft-bodied animals, with which this book is mainly concerned. Second are those of arthropods and vertebrates which possess a hard external or internal skeleton. That two such different groups of animals have adapted themselves independently and successfully for aquatic, terrestrial and aerial life may be associated with the organisation of their locomotory equipment as a series of jointed levers. Their hard skeletons act as the basis for muscular antagonism. The active phase of muscles is in contraction; they are incapable of active elongation, and an external force must be applied to restore them to their initial length. This antagonism may take place between flexor and extensor muscles, as in a limb, or by a hydrostatic or fluid skeleton. This is the prime function of a skeleton, although it also provides support and protection as, for example, in the head of vertebrates.

Whereas in arthropods and vertebrates the same basic design of propulsive mechanism is modified to a wide range of environments, soft-bodied animals are predominantly aquatic in habit and use hydraulic mechanisms for propulsion. These have evolved in different ways in various groups of animals so that the forces developed by a fluid-muscle system subserve the locomotory function in a variety of aquatic habitats, such as burrowing into soft substrates, movement over or into hard material or free swimming.

PRINCIPLES OF LOCOMOTION

The locomotory machinery of animals may be divided into three parts, the engine, transmission and propellor. In the engine chemical energy is transformed into mechanical energy, as in the contraction of muscle. This energy is transmitted to the propellor by a system of limbs acting as levers or, in soft bodied animals, by the fluid of the hydraulic system. The surface of the animal which exerts a force against the environment acts as the propellor. This is commonly a specialised structure in vertebrates, but in primitive metazoans may be represented by a large part of the body wall.

Despite great variation in animal structure, environment and pattern of locomotion, the principles involved in animal locomotion are basically few. They concern Newton's Laws of motion and, for soft-bodied animals, the principles of hydraulics. Newton's Laws may be conveniently expressed in biological terms.[2,43] According to the 1st law, if an animal is at rest relative to its environment then it can only be set in motion by the application of an external force. Consequently, if an animal can move its body by its own unaided efforts it must be able to elicit a force from the external environment by the action of its own propellor.

A force similarly derived from the environment must be brought into play for any change in direction or speed. The application of an unbalanced force (F) to an animal of mass (m) results in acceleration (a) of the animal in the direction of the force. Acceleration is proportional to the force and inversely proportional to the mass of the body. This second law is used to define the unit of force; a Newton (N) is the force needed to give a mass of 1 kg an acceleration of 1 m s^{-2}, i.e. to increase its velocity by 1 m s^{-1} every second. A force of F N gives a body of mass m kg an acceleration of a m s^{-2} where

$$F = ma$$

Since the acceleration of a body is the velocity (V) gained per unit time (t) this equation may be expressed as

$$Vm = Ft$$

Thus the velocity imparted to an animal is directly proportional to the magnitude of the force and the time during which it is applied and inversely proportional to the mass of the body. The second equation may be expressed simply as momentum (Vm) equals impulse (Ft). A convenient example of this may be taken from the analysis of jet propulsion in squid (Chapter 6), where motion is produced by a jet of water from the mantle cavity. This jet may be considered to represent the impulse (Fig. 1.1). With a constant body weight a greater velocity will result from a more powerful jet, while conversely a small squid can achieve the same velocity as a larger animal with a jet of less force or shorter duration.

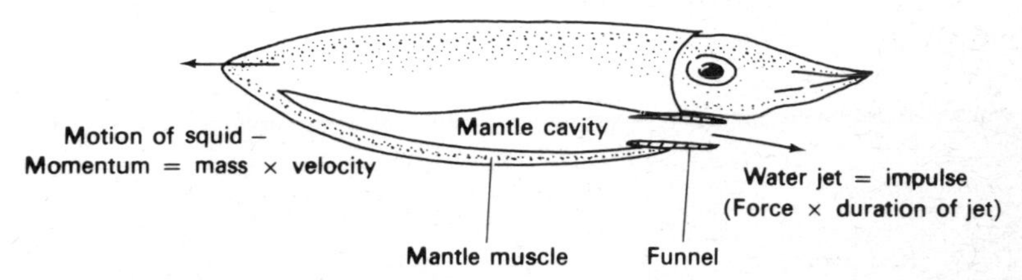

Fig. 1.1 Diagram of a squid sectioned so as to show the mantle cavity and funnel. The pressure generated by the mantle muscle produces a water jet through the funnel and motion of the squid in the opposite direction, the momentum being directly proportional to the impulse.

Newton's third law states that for every action there must be an equal and opposite reaction. In biological terms this means that to thrust forwards as, for instance, in burrowing, an animal must simultaneously exert a backward force against the environment. The burrower only moves forward when the substrate resists the backward movement or in other words provides an anchorage for part of the body. If a worm were to thrust downward with its proboscis without any anchorage of the body, then movement backwards out of the burrow would result. (See Chapter 3).

Motion of any animal requires tension to be exerted in muscles, but the situation is so complex that even in the case of a simple limb it has scarcely proved possible to analyse completely the forces involved. But the overall propulsive force may be determined, since this is equal to the resistance which the ground exerts. It can be measured by allowing the animal to apply thrust to a balanced platform to which an electronic force transducer it attached. The development of electronic recording techniques in recent years, in particular the Statham pressure transducer,[52] has greatly facilitated studies of locomotion of soft-bodied animals.

Another experimental approach is by use of photography to ascertain the duration of the application of force and the velocity of an animal of known weight. An example of this type of analysis may be given in respect of the squid, *Loligo vulgaris*,[86] whose motion was analysed by cine and stroboscopic photography which allowed instantaneous velocity curves to be drawn. The rapid expulsion of water from the mantle cavity in the form of a jet propels a 0.1 kg squid along the line of application of the force with a maximum acceleration of about 30 m s^{-2}. The force is equal to the rate of change of the momentum of the squid, namely $0.1 \times 30 = 3$ N. When comparing the locomotion of different animals it is often necessary to consider the work carried out during specific movements. Work, measured in joules, is done when the point of application of a force is moved and may be determined by multiplying the distance moved by the force or by its resolved component in the direction of motion. Thus if the squid moves directly along the axis of the jet for a distance of 1 m, the work done would be 3 joules. Power is the rate of doing work, and for the period of acceleration caused by the production of a single jet of 100 m s duration would be 30 watts (= J s^{-1}).

HYDRAULIC SYSTEMS

Soft-bodied animals utilize the fluid skeleton both as a mechanism of locomotion and for support. The simplest condition that may be considered as an example is that of a long, cylindrical worm, somewhat similar to the form of *Arenicola*, which consists in its essentials of a flexible muscular body wall enclosing an incompressible liquid medium (Fig. 1.2). The muscles are organized in circular and longitudinal layers and each group functions antagonistically to the other by means of pressure being applied to the contained fluid. Thus contraction of the circular muscles reduces the diameter and causes extension of the worm. Recovery is only possible by contraction of the longitudinal muscles, the circular muscles being stretched by the pressure generated in the fluid. Where fluid is used as a basis for muscular antagonism we have a hydrostatic or fluid skeleton.[22] When a liquid is used as a skeletal element it must conform to the mechanical requirements of being incompressible, readily deformable and of constant volume. These properties are well fulfilled by water, which is virtually incompressible and has low viscosity, whilst the constancy of volume is inherent in the animal design.

Local contractions of circular muscle can cause more complex changes of shape, for instance the extension or thickening of other

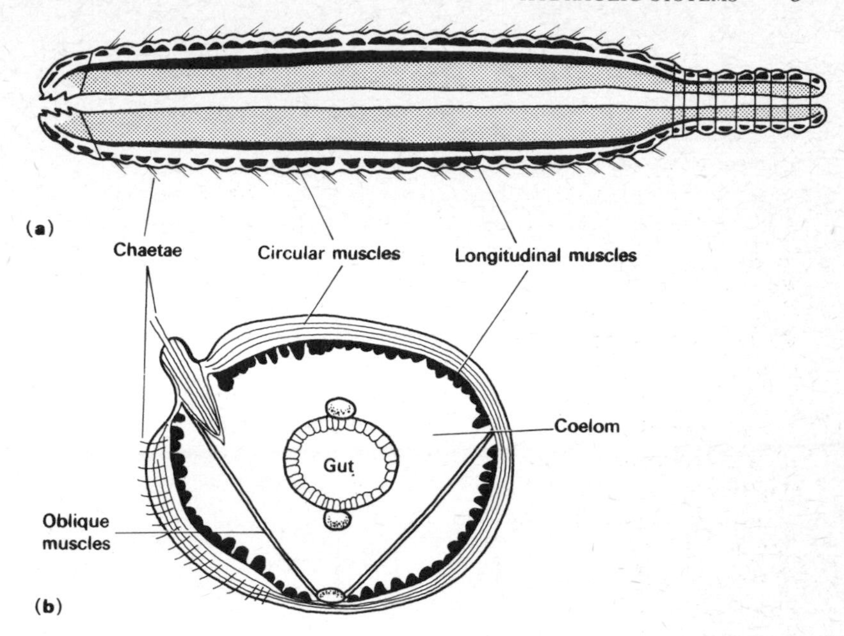

Fig. 1.2 Diagrams of *Arenicola* in longitudinal (**a**) and transverse (**b**) section indicating the extent of the single trunk coelom (stipple in **a**), the septate tail region and the circular and longitudinal musculature. The transverse section shows the location of notopodial (above) and neuropodial chaetae only on the left.

parts of a vermiform body (Fig. 1.3), provided that appropriate muscles are relaxed. Restoration of an elongated form (Fig. 1.3b or c) can only be brought about by contraction of the longitudinal muscles, but if the length of the body is held constant (d), then contraction of the circular muscles in the dilated region, and their relaxation at the right end of the body, would restore the worm to the original condition. Contractions of any muscles about a hydrostatic skeleton produce a pressure in the fluid which is transmitted in all directions. Provided the length of a vermiform body is constant, then any pressure developed is applied generally to the body wall and tends to stretch all muscles, restitution generally being dependent on the controlled relaxation of specific circular muscles. Thus a resting pressure of about 2×10^{-2} N cm^{-2} water pressure recorded in the trunk coelom of *Arenicola* must stretch the circular muscles when they are relaxed after the passage of each peristaltic wave (p. 56).

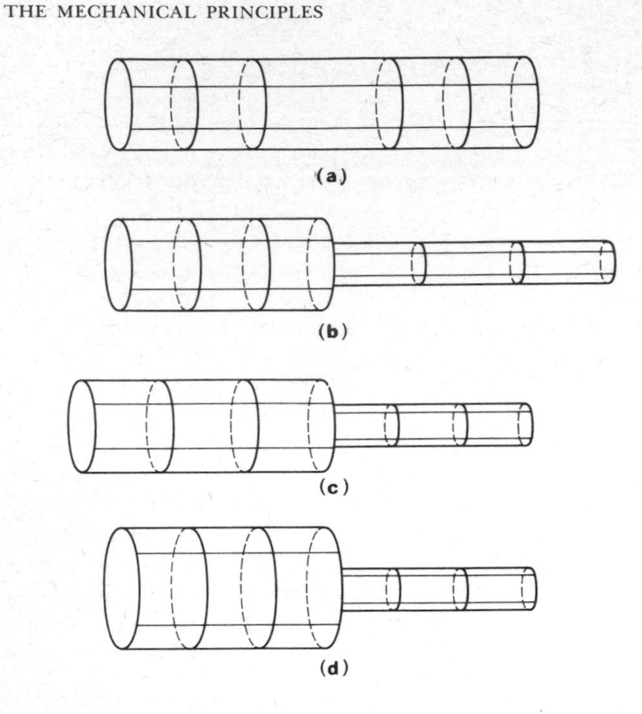

Fig. 1.3 Consequences of the contraction of the circular muscles of the right half of a cylindrical worm. (**a**) initial state with all muscles relaxed; (**b**) left half maintained constant by muscle tension, extension of right; (**c**) and (**d**) right side remaining at constant length, extension of left at constant diameter (**c**), or increase of diameter at constant length (**d**). Restoration to the initial state occurs by contraction of longitudinal muscles in b and c whereas d requires contraction of circular muscles (left).

Other orientation of the body wall musculature to serve the same function is mechanically possible. If fibres are randomly arranged the truly circular muscles would antagonize all others to various extents according to their orientation. However, such an arrangement would place considerable demands on the nervous system, which is at its simplest with the muscles lying in two planes, for then excitation and inhibition of the two sets of muscles on a segmental basis is all that is required.

When fluid is used as a skeletal element it must conform to the mechanical requirements of being incompressible and readily deformed. The hydrostatic skeleton does not necessarily have to contain fluid which can flow freely, however. A parenchymatous tissue loosely constructed, as in nemertine worms, may serve, for it may be sufficiently deformable to allow small changes in shape and to permit muscular antagonism. But, wherever fluid displacement is required to affect a major change of shape, as in the swelling of the body wall of *Arenicola* (p. 56) or eversion of the proboscis of *Priapulus*, then the animal must possess a liquid-filled cavity of sufficient dimensions to allow fluid transfer. Animals in which the change of body shape is greatest generally exhibit the largest liquid-filled body cavities. In this sense a body cavity containing a liquid medium may function as a hydraulic system. In such a system the force derived from muscle tension may be transferred to another region of the body where the pressure may be utilized, for instance in locomotion. This may also involve fluid flow. Hydraulic systems thus include hydrostatic skeletons by definition, but extend beyond them in so far as force generated may not be utilized simply to antagonize a complementary set of muscles. Pressure generated by all or part of the body wall musculature can do work in another region, for example in the forcible eversion of the proboscis of *Sipunculus* (p. 72). However, one disadvantage of the hydraulic transfer of force in animals with a single hydraulic system is that there can be no movement of one part without causing some disturbance to the rest of the body.

A hydraulic system has one additional requisite, constancy of volume of the contained fluid. This presents little difficulty in animals with parenchymatous tissue such as nemertines, for free liquid is minimal; but to a coelomate animal with a fluid-filled cavity, open nephridia or coelomoducts may present a hazard. However, these segmental organs are provided with sphincter muscles which control the loss of body fluid. In *Arenicola*, which has a large fluid-filled trunk coelom, these sphincters are a specialized part of the circular muscles of the body wall and can control the loss of coelomic fluid at pressures of up to 1.5 N cm^{-2}.[24]

In the earthworm the coelom is divided by septa into a large number of segments and, as we will see later (Chapters 3 and 5), studies of the hydraulic systems of worms shed considerable light on interpreting the origin and significance of metamerism. The septa are incomplete, each having a foramen around the nerve cord, but there is a sphincter muscle which probably closes tightly and makes the septum water-tight. Thus the antagonistic circular and longitudinal muscles of each segment may operate independently with the segmental volume constant, and pressure can be different in different parts of the body. Although the septa must remain taut under pressure for segments to act independently, it is probable that each cannot sustain by itself any considerable pressure difference, but rather that a succession of septa may maintain a pressure gradient. This has been shown in experiments recording the pressure in single segments of different regions of the earthworm.[98] In such a system, while each segment may operate largely independently in extension and contraction, force cannot be transferred from one region of the body to another nor can there be any large fluid displacement. Thus in the septate condition, the effective mass of the body cavity functioning as a hydraulic system is greatly reduced. *Arenicola* shows both of these conditions, for the large continuous trunk coelom, in which high pressures have been recorded during burrowing (p. 56), functions as a hydraulic system in so far as the force of muscle contraction posteriorly may contribute to dilation of the body wall anteriorly, and as a skeletal system whereby the intrinsic circular and longitudinal muscles may antagonize each other. The high pressures of the trunk coelom are not found in the tail, whose segments are effectively isolated from the trunk and from each other by septa (Fig. 1.2a). In the tail the coelomic fluid would appear to have only a skeletal role for the antagonism of the longitudinal muscles during defaecation.

THE DETERMINATION OF PRESSURE

Brief consideration of the generation and measurement of pressure is appropriate in view of its importance as the effective mechanism in hydraulic systems and its value as a measure of the work potential of an animal. Pressure may be defined as the force which would act on a unit area either side of a plane surface at a given position in a fluid. By this definition, pressure is a force per unit area and can be expressed as N cm^{-2} or, in the case of large pressures, in terms of atmospheric pressure at sea level.

A basic method of measuring pressure is with a manometer. This is a U-tube partly filled with liquid, one arm being connected to the animal under investigation and the other to the atmosphere. The pressure difference between the animal and the environment is represented by the difference in the height of the liquid in the arms. This height is dependent on the density of the liquid, water or mercury being commonly used. Pressure may thus be expressed in cm of water, 1 cm water being equivalent to about 10^{-2} N cm^{-2}. For most experimental purposes manometers are unsatisfactory. They do not work properly if the arms are fine tubes, because of the effects of surface tension, while with wider arms the inertia of the mass of liquid may prevent the manometer from following rapid changes in pressure. Additionally the volume displaced would represent a great proportion of the body fluid of the animal being examined. Any increase of pressure in the coelom of *Arenicola*, for example, would result in an important proportion of the coelomic fluid being lost to the manometer, so preventing the system from operating at constant volume and having the effect of reducing the pressure. An electronic pressure transducer, such as the Statham transducer, is much more satisfactory. It is essentially a small thick-walled container with a metal diaphragm at one side. Increased pressure deforms this and activates a strain gauge mounted just beyond the diaphragm. Pressure fluctuations cause a negligible change in the volume contained in transducers; e.g. 0.0024 per 1 N cm^{-2} in the P23 BB Statham Transducer, which has commonly been used by zoologists.

If the body fluid of a worm is at different pressure, P, from its surroundings, then a force, P/unit area, will act on all the body wall. If part of this is a curved surface, possibly dome-shaped, the force will act over the whole surface but in different directions in different parts since it always acts at right angles to the surface. The resultant outward force is not P times the area of the curved surface but P times the plan or projected surface area. If the body of a worm is thought of as a tube of radius r, containing liquid, and with muscles in the walls, then, if it is cut by a longitudinal horizontal incision, the pressure (P) would tend to force the two halves apart (Fig. 1.4). If the length of the worm is l, the projected area of each half is $2rl$ and the force would be $2rlP$. This can be expressed as a tension per unit length of the body wall acting around the circumference, i.e. in the plane of the circular muscles. The total length of the cut edges of each half of the worm is $2l$ so that the tension T_c (N cm^{-1}) is $2rlP/2l$ or $P = T_c/r$. If the thickness of the circular muscles is known, then the

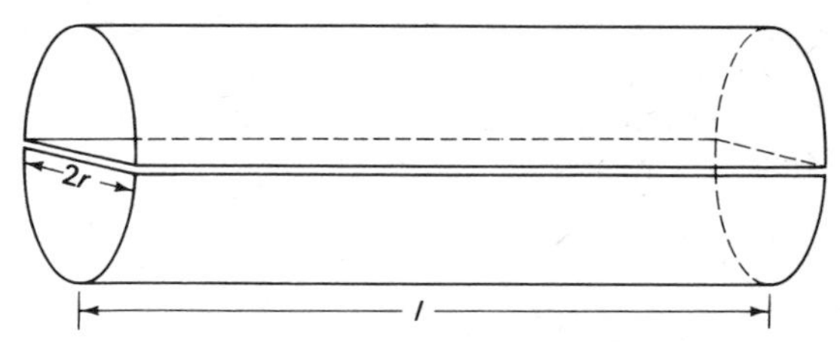

Fig. 1.4 Diagram of part of the cylindrical body of an unsegmented coelomate worm cut by a horizontal longitudinal median incision. This is used as a model for the determination of the relation between fluid pressure and tension in longitudinal and circular muscles. *l*, length of body; *r*, radius.

tensile strength exerted to produce a known pressure may be determined in terms of their cross-sectional area (N cm^{-2}). From the trunk coelom of the lugworm *Arenicola marina* a pressure pulse of 1 N cm^{-2} has been commonly recorded. The worms had a trunk of mean radius 0.6 cm, so that the tension (T_c) would be 0.6 N cm^{-1}. A longitudinal section of the circular muscles of the body wall 1 cm in length has an average thickness of 0.02 cm. Thus these muscles must exert a tensile strength of 30 N cm^{-2} ($0.6/0.02 \times l$) to produce such a pressure pulse.

If the worm is thought of as a tube with ends then there will be forces of $\pi r^2 P$ (area of end × pressure) tending to move them apart. These will cause longitudinal tension (T_L) in the body wall, equal to the force divided by its circumference or $P = 2T_L/r$.

THE SKELETON OF SEA ANEMONES

One of the simplest examples illustrating the principle of hydrostatic skeletons is the sea anemone, *Metridium*, which is essentially a cylinder of very variable shape containing sea water (Fig. 1.5). Unlike most hydraulic systems the liquid-filled body cavity, the

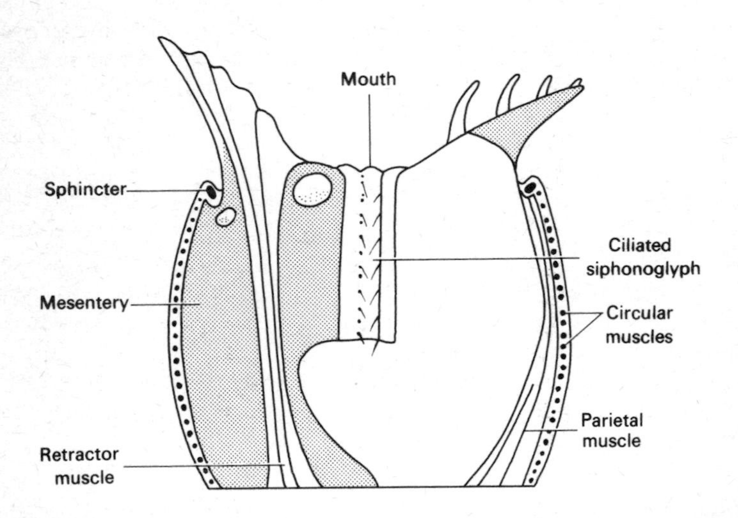

Fig. 1.5 Diagram of a vertical section through the sea anemone, *Metridium* showing the arrangement of muscles concerned with changes in shape and size. (after Batham and Pantin[11])

coelenteron, is not permanently closed but opens to the exterior by means of the stomodaeum. Parts of this are ciliated, forming the paired siphonoglyphs, by means of which water is drawn into the polyp. From the inner wall of the cylinder radially arranged muscular flaps, termed mesenteries, extend inwards from the wall of the cylinder. Both body wall and mesenteries consist of a jelly-like substance, mesoglea, sandwiched between two layers of cells. These structures allow the anemone to make extremely large changes of shape ranging from complete contraction where it appears shrivelled, to considerable distension, an increase in size of five or six times. During this expansion water is taken in through the ciliated siphonoglyph. The majority of sea anemones, when expanded, present to the eye a nearly constant shape, but close and prolonged inspection, or recording such as carried out by Batham and Pantin,[11] shows that the shape is changing very slowly as the result of continual activity. In *Metridium*, and probably most other actinians, the coelenteric contents act as a hydrostatic skeleton with little uncontrolled loss of water through the mouth, the stomodaeum acting as a sleeve valve preventing outflow, so that a constant volume may be maintained during these changes of shape. The fluid-muscle system has circular muscles running round the body wall like the hoops of a barrel which are naturally antagonistic with the longitudinal muscles in the mesenteries (Fig. 1.5). The latter include the retractor and parietal muscles. All these muscles are characteristically very extensible and may stretch from 200 μm to 1–2 mm in length. The limits of changes of shape of a tubular animal are set by the elasticity of the body wall and the operating length of the muscles. Contraction of the column from the expanded condition leads to buckling of the body wall and reduction of volume, but antagonism between the two sets of muscles continues so that reduction of diameter still causes an increase in length and *vice versa*.

Batham and Pantin[11] made numerous recordings of the pressure within the coelenteron and related it to activity (Fig. 1.6). These recordings show that the average pressure is very low, about $2–3 \times 10^{-3}$ N cm^{-1}, and that this fluctuates as the anemone continually makes small movements. Part of this resting pressure must be used to support the polyp, particularly when it is expanded vertically. When freely suspended in water the weight of a specimen of *Metridium*,, 5 cm in diameter, was found by Batham and Pantin to be 0.76 g. Some of this weight will not be supported by the coelenteric pressure but assuming that all of it is, and that the body is completely flexible, then the pressure required to support the weight alone would be about 0.4×10^{-3} N cm^{-2} (wt of anemone/cross-sectional

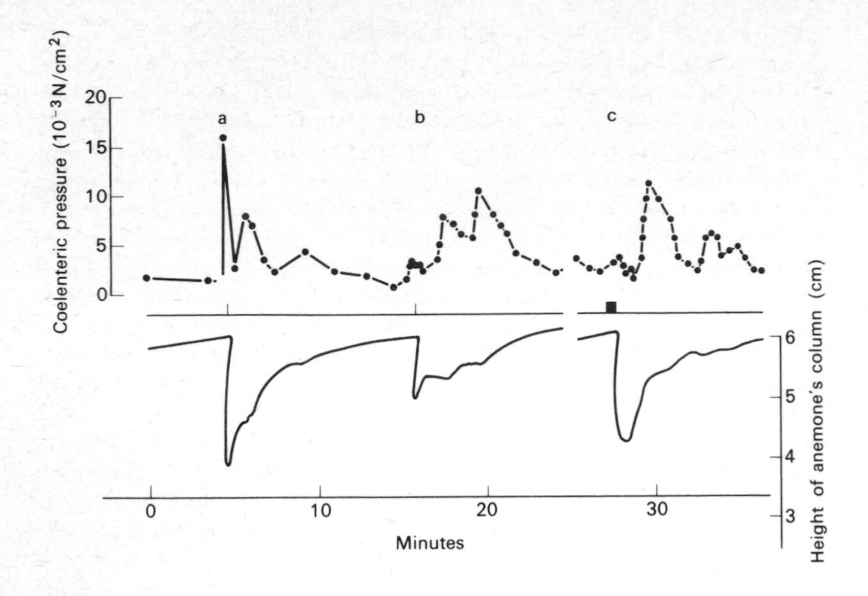

Fig. 1.6 Pressure changes (*above*) and contraction of body wall of anemone (*below*) following electrical stimulation. (**a**) retractor response to three condenser shocks at 1 s intervals. (**b**) retractor response to two condenser shocks at 1 s intervals. Note small rise in pressure accompanying single retraction and large rises in pressure accompanying re-extension. (**c**) ten shocks at frequency of 3.2 s elicit parietal contraction. This causes no significant rise in pressure. Subsequent re-extension accompanied by large rises in pressure. (after Batham and Pantin[11])

area). Although this pressure is minute, it is not negligible compared with the actual coelenteric pressure. Indeed, at times of full expansion the anemone may not be able to support its own weight and it then falls over to one side. Electrical stimulation of the parietal muscles caused contractions of the column and, as in Figure 1.6(a), a brief pulse of higher pressure; the absence of a high pressure with contraction in the subsequent recordings (b and c) is possibly due to the manometric technique of pressure recording used by these authors. The important feature is the sustained rise of pressure accompanying re-extension, associated with contraction of the circular muscles. Thus extension of the column after parietal contraction is an active process due to the antagonistic contraction of the circular muscles.

Besides the fluid-muscle system, an anemone also has the support of mesoglea. If a *Metridium* is narcotized with magnesium chloride solution so as to put the muscles out of action it will gradually acquire a median size and shape. This presumably represents the form in which the mesoglea is unstrained. When the anemone is fully distended the mesoglea is stretched, whereas when the animal is contracted the mesoglea is compressed and thickened. From either of these two extremes the mesoglea will tend to return to its unstrained condition. Thus at maximal size the resting pressure of 2×10^{-3} N cm^{-2} in the hydraulic system could be considered as antagonizing the mesoglea and maintaining it in the stretched condition. When an anemone is poked it generally contracts fully in a few seconds because of the strength of the retractor muscles. Reinflation would take more than an hour because of the high viscosity of the mesoglea and the need to take water into the coelenteron. This is accomplished slowly by means of the ciliated siphonoglyph. High viscosity means that it is harder for an anemone to stretch rapidly than slowly, a condition which allows hard knocks to be withstood, as by a wave, but allows low sustained pressures to cause expansion. Initial enlargement from full contraction may be due to ciliary currents drawing water into the anemone, supplemented by slow suction as the compressed and thickened mesoglea relaxes into an unstrained condition. Batham and Pantin[11] reported short pulses of negative pressure (of about 2×10^{-3} N cm^{-2}) during expansion, which could be explained as periods of suction caused by the mesoglea. During expansion the coelenteron may be closed to the exterior for a short time, allowing the anemone to make changes in shape by muscular antagonism. The absolute volume of water contained in the body is not critical, provided the volume is constant at any instant. Functioning at small size can only mean that the muscles must be capable of contraction at very reduced lengths.

The functioning of anemones illustrates how a gut can operate as a fluid skeleton. However, it has two principal disadvantages. First, the mouth must be capable of being closed tightly during changes of shape; second, the use of the gastric cavity as a hydrostatic skeleton prevents specialization of the gut for extracellular digestion. As parts of the body wall contract, so the liquid moves freely within the coelenteron, and this prevents an orderly sequence of digestive enzymes being presented to the food. Anemones have no anus and digestion is largely intracellular. Isolation of the gut from the body cavity, as occurs in the annelid worms, separates the alimentary and hydraulic systems, and permits much greater specialization of the digestive tract.

THE HYDRAULIC SYSTEM OF ASCIDIANS

Observations on the hydraulic system of *Ascidia* permit comparison with *Metridium*. Ascidians contract and eject a stream of water at fairly regular intervals. This activity was monitored by inserting a fine plastic cannula into the water-filled branchial chamber and attaching it to a pressure transducer. Pairs of pin electrodes placed superficially in the test of an animal adjacent to the branchial chamber (Fig. 1.7d) were attached to an impedance pneumograph so that any change in the dimensions of the test affected the electrical impedance between the electrodes. These signals were observed on a pen recorder simultaneously with pressure.[40] Contraction of the mantle muscles results in reduction of internal volume and the emission of a jet of water with an associated pressure pulse (Fig. 1.7a). Contraction takes place in both the longitudinal and transverse directions simultaneously, so that muscles in these places cannot act antagonistically to each other. Restoration of the animal

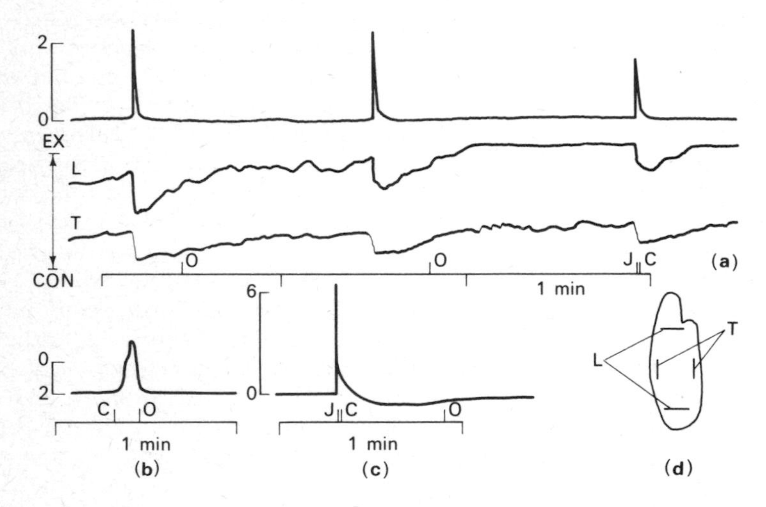

Fig. 1.7 Pressure (10^{-2} N cm^{-2}) and impedance recordings of the activity of ascidians with opening (O) and closing (C) of the siphons and the production of water jets (J) marked. (**a**) *Ascidia nigra*, branchial pressure above, extension (EX) and contraction (CON) of test along longitudinal (L) and transverse (T) axes with electrodes placed as in d; (**b**) *A. nigra* single pressure pulse; (**c**) *A. interrupta*, single pressure pulse showing negative pressure during relaxation with siphons closed; (**d**) diagram of *A. nigra* showing location of electrodes, used in a. (from Goodbody and Trueman[40])

to its initial shape takes place over some 30 s (Fig. 1.7a) and must therefore involve some other mechanism.

It should be stressed that, since muscular contraction produces a jet of water from the branchial chamber, this reduction in internal volume is incompatible with branchial water being used as the basis of a hydrostatic skeleton. Moreover, during at least part of the time of restoration of shape the siphon is closed, so that it is unlikely that ciliary activity reinflates the test. With one species investigated, *Ascidia interrupta*, a negative pressure was commonly observed after the positive pulse (Fig. 1.7c). This only occurs with the siphons closed (compare Fig. 1.7b), which suggests that the negative pressure may be produced by the expansion of the test when water is prevented from entering the body freely. This is comparable to the negative pressures observed in *Metridium*. Restoration of body shape appears to be due to the inherent elasticity of the test being used as an energy store to act antagonistically. Ascidians are thus examples of fluid-muscle systems in which muscles are antagonized by the inherent elasticity of another part of the body. This condition is directly comparable with that obtaining between longitudinal muscle and cuticle in Nematoda, and between adductor muscle and hinge ligament in bivalve molluscs (see Chapters 3 and 5).

2

Movement over Hard Surfaces

CILIARY LOCOMOTION IN FLATWORMS AND NEMERTINES

Two principal effector mechanisms, cilia and muscle, are utilised in movement over hard surfaces by free living flatworms. Parasitic members of the phylum Platyhelminthes are modified in structure and here we are only concerned with the locomotion of the free-living group, the Turbellaria. As one of the more primitive metazoan groups their structure is simple. They are triploblastic animals in which the mesoderm has allowed some development of organs. Their basic structure is illustrated by a section (Fig. 2.1a), showing the orientation of muscle layers and the mesodermal parenchyma. This is commonly recognized as a syncitium with the interstices filled with fluid. There is no discrete body cavity and the parenchyma serves as the fluid in the worm's hydraulic system. The parenchyma is clearly of constant volume and incompressible, but does not flow freely and must therefore be a limiting factor in relation to change of shape in this phylum of worms. It would appear that in flatworms and in certain other animals, e.g. *Ascaris* and squid mantle (p. 142), tissues rather than freely displaceable fluids suffice for force transference, at least locally, between antagonistic muscles of the hydrostatic skeleton. But a fluid of low viscosity is necessary for the transfer of force over long distances or for major changes in shape. It is notable that the animals which burrow most powerfully, generating the greatest pressures, are

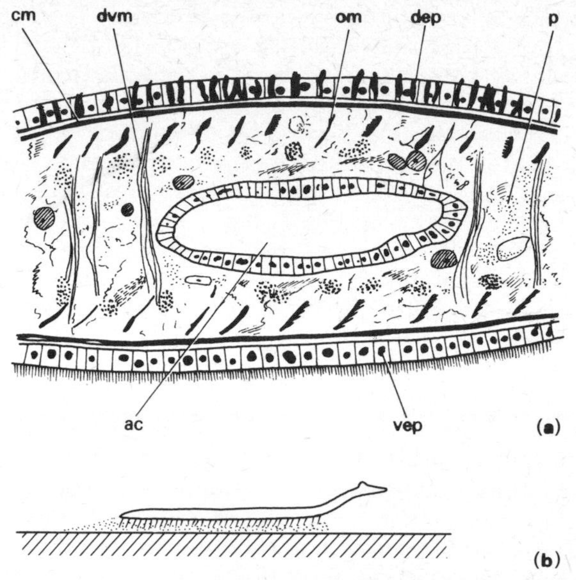

Fig. 2.1 (**a**) Section of planarian showing organization of muscles and parenchymatous tissue. ac, alimentary canal; cm, circular muscle; d ep, dorsal epithelium with rhabdites; d vm, dorso-ventral muscle; om, oblique muscle; p, parenchyma; vep, ventral ciliated epithelium. (**b**) representation of a planarian moving forwards with head raised and cilia beating in mucus (stipple). (after Pearl[91])

those with the largest liquid-filled cavities. In the Turbellaria, the cellular parenchyma is sufficiently plastic to permit some limited changes of body shape. In the Nemertea, which are of the same grade of tissue organization, the parenchyma is more gelatinous, nearly approximating to a fluid skeleton, and this allows the use of peristaltic waves as a means of locomotion. Moreover, nemertine worms possess a large fluid-filled cavity, the rhynchocoel, which makes considerable changes in shape possible when everting the proboscis.

Both Turbellaria and Nemertea are largely held in shape by collagen-like fibres which forms a trellis in the basement membrane of the surface epithelium. The manner in which these investing fibres function in worms of different shape is discussed by Clark[26] and Alexander.[2]

The epidermis of turbellarians and nemertines is commonly heavily ciliated, many species using the cilia for movement. Ciliary mechanisms are not subject to fatigue like muscles, but can only provide sufficient locomotory power for animals of small mass. Thus acoel and rhabdocoel worms, which are small, move almost exclusively by cilia, whereas larger triclads and polyclads show a greater variety of structure and locomotory habit. Aquatic triclads, e.g. *Planaria*,[91] crawl over the substrate by cilia. These are particularly well developed on the ventral surface, where a copious secretion of mucus attaches the animal to the substrate and is the medium in which the cilia beat (Fig. 2.1b). The characteristically flattened shape of Turbellaria is related to absence of respiratory or circulatory systems and the consequent need to reduce the distance through which oxygen must diffuse to the tissues.[3] The propulsive force of the planarian is, however, proportionate to the surface area, and the flattened shape is thus a locomotory adaptation as well as a respiratory one. A specimen of *Planaria maculata* 11 mm long can crawl at about 0.15 cm s^{-1} compared with 0.12 cm s^{-1} for a specimen of the same species 6 mm in length. Thus the increase in surface area has rather more than compensated for the increase in mass. In terrestrial triclads, however, a nearly circular cross-section is maintained so as to present minimum surface area to the air to reduce water loss. Consequently they present a small propellor surface to the substratum in relation to their mass and indeed the ciliation of the ventral surface is somewhat restricted. Life in air rather than in water also increases the effective mass of the animal so that the cilia are required to perform more work. It is hardly surprising that locomotion of terrestrial planarians is slow and that they commonly move by muscular contraction rather than by cilia.

Nemertines characteristically have highly extensible bodies. For example the body of *Lineus* may be some 10 times longer when the circular muscles are fully contracted than when they are relaxed. Although not characteristically of flattened shape, nemertines may nevertheless progress at slow speeds by means of cilia with no visible signs of muscular locomotory waves. As in flatworms, mucus is always secreted in association with ciliary locomotion, for it is against mucus that the cilia exert their propulsive effort. The terrestrial nemertine, *Geonemertes*, secretes a tube of mucus through which it is propelled by cilia, apparently only using its musculature to change direction.[89]

PATTERNS OF MUSCULAR ACTIVITY

When an animal moves through a fluid both inertia and viscosity must be considered, the ratio between inertial and viscous forces being expressed as the Reynolds number (R = $LV\rho/\mu$ where L is the characteristic length of the system, V, velocity, ρ, density of medium and μ its viscosity). In microscopic organisms inertial forces are negligible, viscous forces being dominant, but where the size increases to more than a few millimetres inertial forces become more important and greater forces are required for locomotion.[23] Turbellaria range from less than 0.1 to about 5 cm in length. The smallest are thus comparable in size to ciliates and move about in much the same manner, but the larger almost exclusively utilize muscular contraction for locomotion and can thus achieve greater speeds, e.g. up to 0.5 cm s^{-1} by the aquatic polyclad *Leptoplana*. Indeed, most polyclads move faster by muscular locomotion than do aquatic triclads by cilia. It may be noted that planarians, such as *Dendrocoelum*, may, under certain conditions, temporarily abandon ciliary locomotion and move very much more rapidly with long muscular steps. The transition from ciliary movement to muscular locomotion over hard substrates can be conveniently studied in the Turbellaria, where it marks the origins of the dynamic functioning of the fluid-muscle system of triploblastic animals and illustrates the application of muscular force without the systems of levers developed in animals with hard skeletons.

Three distinct patterns of muscular locomotory activity, namely looping movements, pedal locomotory waves and peristalsis, have developed in Turbellaria and Nemertea, albeit in a rudimentary and often imperfect fashion. These techniques are commonly more effectively developed in other animals and will be further discussed in relation to these.

Looping

Pearl[91] observed that when *Planaria* is stimulated posteriorly, the tail is protracted and the head extended. The anterior end is attached to the substrate, probably by mucus, and the posterior is drawn up by contraction of longitudinal muscles. The posterior end may then be attached and the head extended. This locomotory cycle may be repeated several times before the animal returns to ciliary locomotion (Fig. 2.2). Two points emerge from this. Firstly, whenever any movement is made then part of the animal is stationary, anchored to the substrate. This is in keeping with Newton's third law, for the worm thrusts forward because the ground resists the backward force from the body, the force derived from the ground being referred to as a static reaction. Clearly the force with which an animal can move forward depends on the

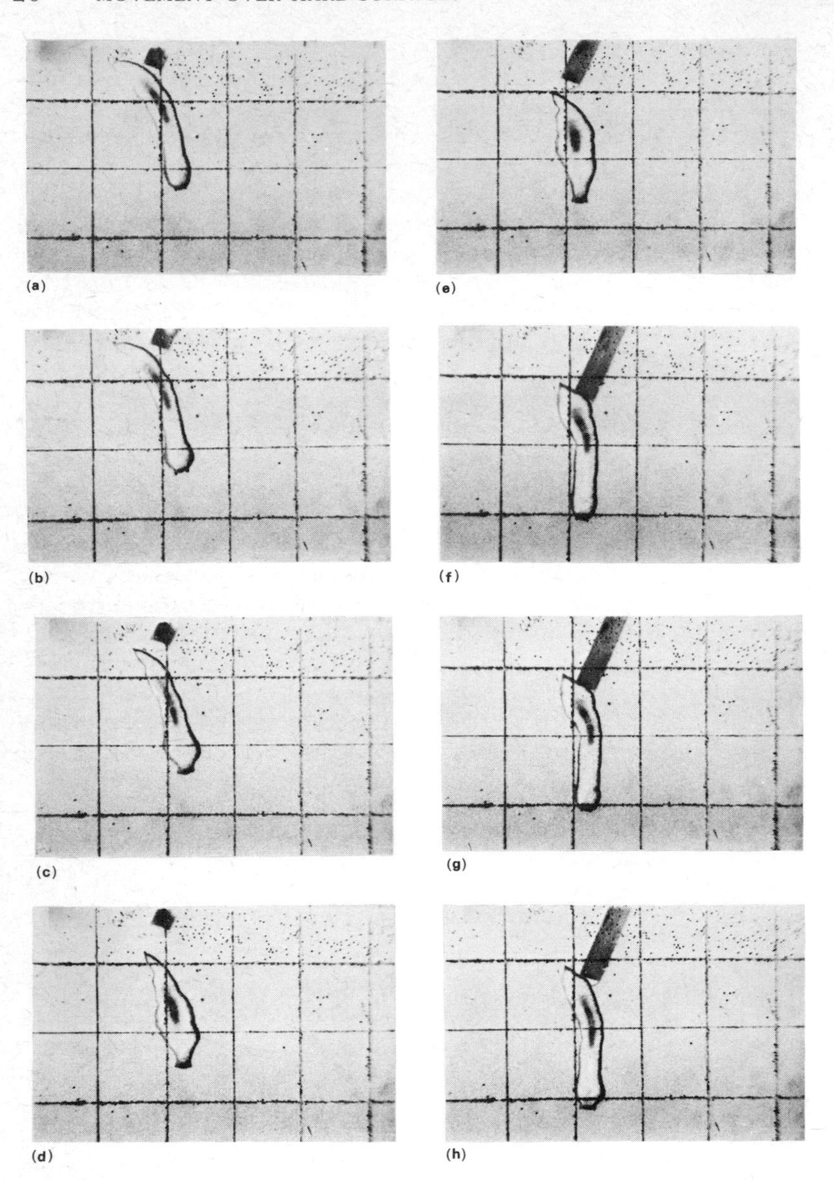

Fig. 2.2 Series of photographs of the planarian, *Dendrocoelum*, taken at intervals of 0.5 s showing muscular locomotory waves produced by contraction of longitudinal muscles upon tactile stimulation (1 cm grid). (photograph Dr H. D. Jones)

effectiveness of the anchorage, so that on a hard surface, with only the frictional resistance of mucus as an anchor, relatively small forces may be exerted. Secondly, extension of the body forwards must involve contraction of circular or transverse muscles, since contraction of the longitudinal muscles is incompatible with protraction of the body. This condition resembles that shown in Figure 2.1b, where the head is raised while the tail effectively acts as an anchor. To accomplish this, some turgor pressure must be developed in the parenchyma. Clark[26] suggests that this may be achieved by contraction of the oblique muscle layer (Fig. 2.1a), although there is no experimental proof of this. Looping locomotion occurs in flatworms only as an escape mechanism.

Malacobdella, a nemertine commensal in the mantle cavity of bivalve molluscs, moves in a comparable manner and has developed a permanent sucker at the posterior end of the body.[32] The sucker is attached (Fig. 2.3a), providing a secure anchorage for extension of the worm forwards by contraction of circular muscles. A sucker is a better anchor than one provided by mucoid adhesion and allows more force to be developed for extension of the worm. The anterior part is then attached, possibly by mucus (Fig. 2.3b), the posterior sucker detached, and the worm is drawn forward by contraction of longitudinal muscles (Fig. 2.3c). This locomotory cycle may then be repeated.

Leeches have developed this type of movement further by development of both anterior and posterior suckers (Fig. 2.3d–f). With

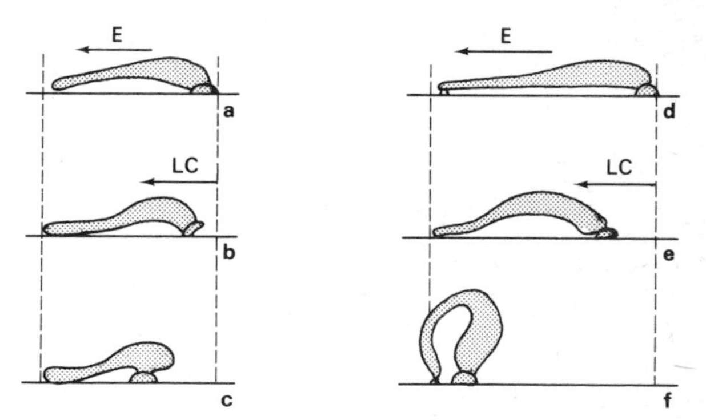

Fig. 2.3 Diagrams illustrating locomotion by looping movements **a, b, c,** in the nemertine, *Malacobdella* (after Eggars[32]) and **d, e, f,** in the leech, *Hirudo*. Similar locomotory stages are shown in each series. E, extension of body by contraction of circular muscles; LC, contraction of longitudinal muscles advancing posterior region of the worms. Broken lines indicate body length for comparison with length of step.

secure anchorage at both extremities, locomotion is more efficiently executed than in nemertines and may involve more complex muscular activity than simple antagonism of longitudinal and circular muscles. While extension is accomplished in the leech simply by contraction of circular muscles, differential contraction of the longitudinal muscles, together with a flattening of the central part of the body, results in flexion and enables the posterior sucker to be attached close to the anterior one.[22] Thus the 'step' length is increased in comparison to a nemertine and almost equals the length of the leech.

Looping in this manner represents a specialized modification of creeping by alternate extension and shortening of the body. Its efficiency depends not only on the organization of longitudinal and circular muscles but also on the effectiveness of anchorage at each extremity. The fluid skeleton is provided by parenchyma in the Turbellaria and Nemertea, while in leeches the space between body wall and gut is largely filled with botryoidal tissue. The fluid of the hydraulic system of all these animals is viscous, yet sufficient change in shape occurs to allow looping movement. An important feature is that the whole body musculature behaves as a single unit. Only when the entire circular musculature has become contracted do the longitudinal muscles commence to contract. Thus the stresses in the hydrostatic system act throughout the body in the same sense at any stage of the locomotory cycle. These stresses must be much smaller than those occurring during peristaltic movements in other worms, where longitudinal and circular muscles exert stress simultaneously in different directions in various regions of the body.[26] In such worms a less viscous body fluid is generally encountered.

Pedal locomotory waves

Some triclads and many polyclads creep by means of a series of waves of contraction that pass along the ventral longitudinal musculature. Thus the cyclical repetition of contraction of circular followed by longitudinal muscles is replaced by the successive contraction of different regions of the longitudinal muscles. A muscular locomotory wave consists essentially of raising a part of the body from the substratum, moving it forwards and reattaching it. This motion places but small demands on the fluid skeleton, and results in a continuous progression over the substrate which can often be mistaken for ciliary movement.

Pedal locomotory waves can be observed in the terrestrial triclads *Geoplana* and *Bipalium*, in which ventrally located longitudinal

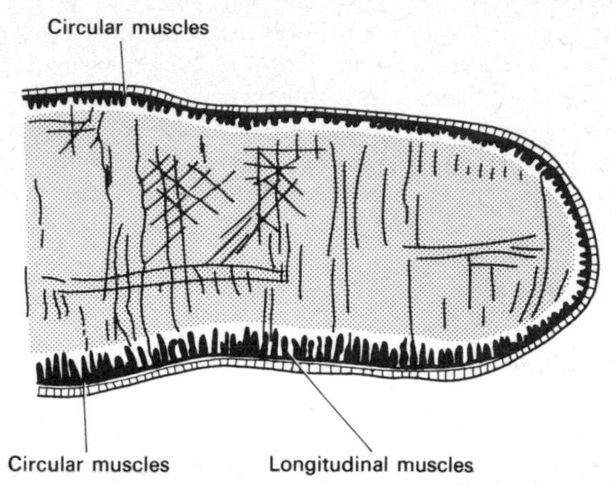

Fig. 2.4 Transverse section of part of the nemertine *Geoplana* showing distribution of circular and longitudinal muscles and the direction of other muscle fibres passing through the parenchyma (stipple). (after Moseley[79])

muscle fibres are particularly well developed (Fig. 2.4). Regions of the longitudinal muscles contract, causing slight local swellings of the body wall and providing points of temporary attachment or anchorage to the substrate (Fig. 2.2). This can be particularly clearly seen[89] in the land triclad *Rhynchodemus* (Fig. 2.5). When the animal moves, large 'myopodia' arise on the ventral surface of the body.

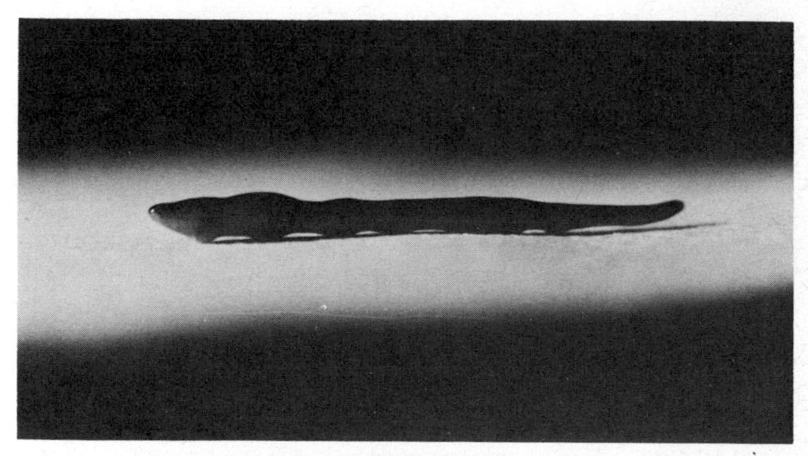

Fig. 2.5 *Rhynchodemus*, a terrestrial planarian, moving to the right using retrograde locomotory waves indicated by the *points d'appui* formed by the body wall. Overall length approximately 1.5 cm. (photograph Dr H. D. Jones)

Several may be observed at the same time, passing backwards along the body; provided the anchors remain fixed to the substrate, then the animal moves forward. Longitudinal muscles are again well developed in this genus, particularly ventrally. The resemblance between such different animals as *Rhynchodemus* and a loping snail such as *Helix* (Fig. 2.19) is remarkable. In both the body is only in contact with the substrate at the extremities of long backwardly travelling muscular waves, and both appear to have a subsidiary propellor, cilia in *Rhynchodemus* and short muscular direct waves in *Helix*. The use of the same rapid propulsive mechanism suggests that this is mechanically the most effective solution to the problem of rapid movement over rigid substrates by soft-bodied animals. Gray[43] further points out that the principles involved are also utilized by such organisms as *Amoeba* and by mammalian fibroblasts.

Peristalsis

Nemertines have two methods of propulsion over rigid surfaces; they may be propelled either by waves of contraction passing over well developed circular and longitudinal muscles as in annelid worms, or by means of cilia which also act as a subsidiary propellor. The rate of movement by cilia is very slow, seldom exceeding 1 cm/min, whereas *Lineus* may travel at ten times this speed by means of muscular contraction. The characteristic form of peristaltic waves may be observed in *Lineus* (Fig. 2.6), where the bulges are produced by the contraction of longitudinal muscles. At any instant these swellings are fixed to the ground and form *points d'appui* against which other parts of the body exert their effort. Peristalsis is relatively poorly co-ordinated in nemertines, so that parts of the

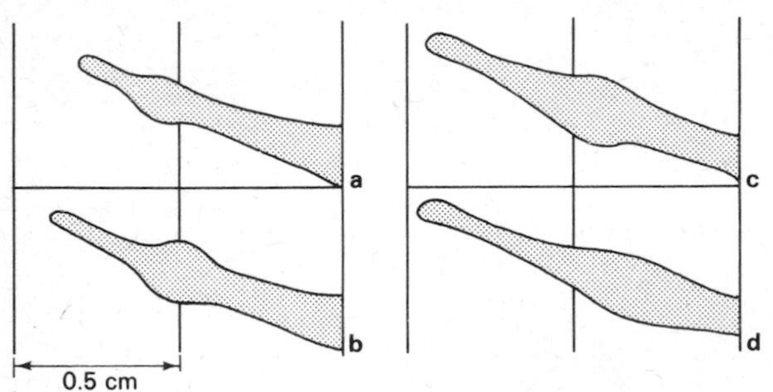

Fig. 2.6 Tracings of 4 successive photgraphs (**a–d**) of a wave passing over the anterior end of the nemertine *Lineus*. Interval between 0.5 s; note change in shape of the wave as it passes posteriorly. (after Gray[43])

body wall of a long worm may show waves which bear little relation with those in other regions. Peristalsis is nevertheless an advanced form of locomotion for this grade of tissue construction, where the largely gelatinous parenchyma acts as a fluid skeleton. It foreshadows peristaltic locomotion with a true fluid skeleton, which is commonly found in coelomate worms and which will be considered below.

THE FORM AND DIRECTION OF PROPAGATED WAVES

It is useful to define at this stage the relationship between movement of an animal's body relative to the ground and the form and direction of locomotory waves propagated over the body. While this is particularly relevant to any discussion of the locomotion of segmented coelemate worms, it is also required for the understanding of the locomotion of snails.

The passage of waves of muscular contraction along a worm-like body to cause forward movement has already been noted in *Geoplana* and *Lineus*, where they pass in the opposite direction to the movement of the worm and are termed 'retrograde waves'. The best known example of this is the earthworm. A locomotory wave moving in the same direction as the body is referred to as a direct wave; an example can be observed in the bodywall of the lugworm, *Arenicola*. These terms, retrograde waves and direct waves, are best defined in relation to the direction of movement of the animal, rather than in respect of head and tail or anterior and posterior, for this simplifies understanding of their mechanisms. However, it may be observed that when an animal moves in a normal manner head first, retrograde waves pass from the head backwards along the body, direct waves run from tail to head.

The passage of locomotory waves is illustrated in Figures 2.7 and 8, where successive regions may be assumed to represent segments of a coelomate worm, but the same argument applies equally well to adjacent parts of the body of a flatworm or nemertine or to the foot of a gastropod. Each segment should be considered to function at constant volume so that it may change in shape (I–VIII) with the contraction and relaxation of longitudinal and circular muscles. The body is considered to be divided up into a number of regions or segments so that each locomotory wave or shortening-lengthening cycle involves eight segments; the phase of contraction of each segment is then one-eighth cycle different from its neighbour. Each wave will cause the body to move a distance equal to the difference

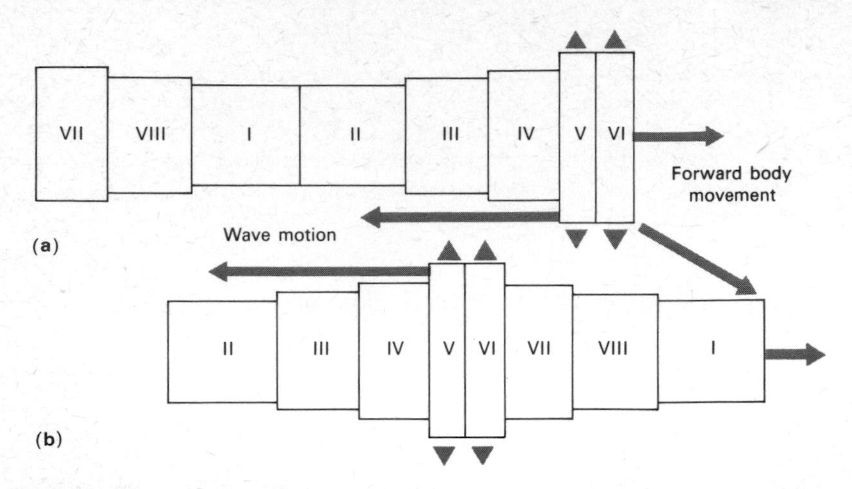

Fig. 2.7 Diagram illustrating the passage of a retrograde wave in a worm. Each segment is $\frac{1}{8}$th cycle out of phase with its neighbours, the phase of contraction being indicated by roman numerals (I–VIII), e.g. maximal longitudinal contraction is shown by V and VI. Segments show approximate length and diameter changes and the arrowheads adjacent to the shortest segments (V and VI) indicate anchorage to the ground.

between the displacement of a wave relative to the ground and its displacement along the body of the animal. The direction of movement of the body depends on two factors: (a) the direction of propagation of the waves and (b) the phase of the contractile cycle at which each part of the body is attached to the ground. Each part is at its shortest and fattest as the longitudinal muscles contract and so may form an anchor or *point d'appui* with the substrate. When the cycle of shortening passes from right to left as in Figure 2.7a and b, i.e. backwards, the region to the right of the anchored segments must lengthen and the animal extends to the right, i.e. forwards. Thus when the fully longitudinally contracted region acts as an anchor the animal is propelled in the opposite direction to the wave, which is therefore termed retrograde. If the segment on the right represents the head, then forward progression entails the movement of the anchored region to the posterior of the animal. On the other hand, if the shortened anchored region were to move from left to right (as from B to A), then the animal is displaced to the left, i.e. in the opposite direction, and so the wave is still retrograde. Indeed it must always remain so, provided that the segments are anchored at their shortest length and consequently at their maximum diameter.

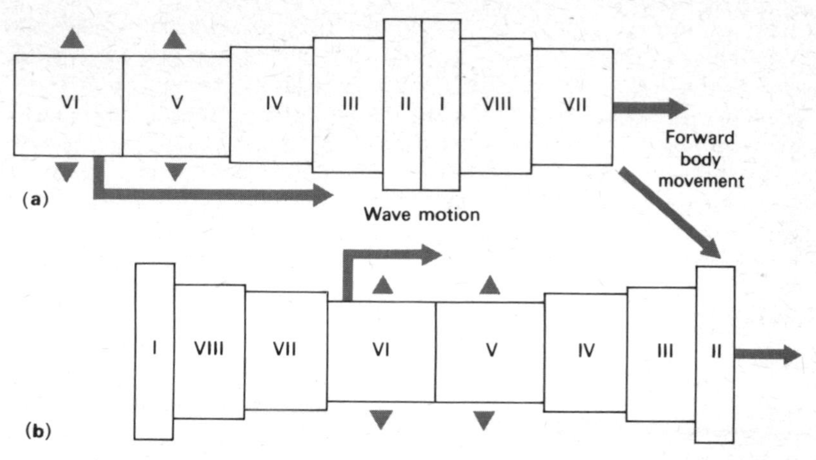

Fig. 2.8 Diagram illustrating the passage of a direct wave in a worm. Details as in Figure 2.5 except that segments now are required to be attached to ground when fully extended (arrowheads).

In contrast with this, forward movement by means of a direct wave, i.e. in the same direction as that of the body, requires segments to be anchored when fully extended. Movement of such a wave, forward from left to right as in Figure 2.8a and b, requires extension of segments prior to anchorage, i.e. in advance of the wave. Thus the body moves forward to the right in the same direction as the wave. A reversal of the direction of the wave (as from B to A), i.e. from right to left, results in movement of the body to the left. If each segment operates at constant volume its extension is accompanied by a reduction in thickness, a condition unsuited to forming a *point d'appui*, particularly in a burrow. Direct waves can only effect locomotion where structural modifications, such as an undivided body cavity, allow maximum extension to be coupled with sufficient thickness for anchorage on the substrate. This occurs in the trunk of the polychaetes *Arenicola* (p. 51) and *Polyphysia* (p. 44) and in the foot of the snail *Helix* (p. 36).

PEDAL CRAWLING IN SNAILS

Introduction

In at least two of the molluscan classes, the Placophora and the Gastropoda, locomotion by a series of muscular pedal waves has been advanced to a high degree compared with the condition of the Turbellaria. In respect of locomotion the gastropod foot is functionally equivalent to the entire turbellarian body, but there is more diversity in gastropod locomotion, from the use of cilia to the development of various complex patterns of muscular pedal waves. Ciliary locomotion is not restricted to the more primitive snails but is seen throughout the class. It is more commonly associated with aquatic snails of small size, e.g. *Nassarius*, and certain pulmonates, in which the relatively weak forces of cilia suffice for locomotion because of their small mass. Speeds attained by snails utilizing ciliary creeping appear comparable to those of flatworms.[26]

Much of our knowledge of locomotion in turbellarians and nemertines is based on visual observations that often date back to the turn of the century. Equivalent descriptions are available for gastropods but, because of the greater ease of carrying out experiments on snails, our understanding of the biomechanics of muscular locomotory waves has been considerably advanced in recent years, for example by the investigations of Lissmann,[65,66] Jones[58] and Jones and Trueman.[59] The types of pedal muscular waves are conveniently summarized by Clark.[26] Here we need only to differentiate between direct waves, as in most pulmonates, and retrograde waves, as in *Patella* and chitons, and also between monotaxic and ditaxic waves. Both direct and retrograde waves may extend right across the width of the foot; these are monotaxic waves, e.g. *Helix*. Alternatively, they may form two parallel systems, when they are termed ditaxic waves, for example in *Patella*. An advantage of the differentiation of the foot into separate functional regions is the ability of the limpet to turn around without moving forwards, for in this genus retrograde locomotory waves may move in opposite directions on either side of the foot.[59] This causes rotation in a similar manner to the simultaneous pulling and pushing on the oars of a rowing dinghy. In *Pomatias* the lateral regions of the foot are more clearly differentiated, one side moving forwards while the other serves as an anchor attached to the ground.[43] The whole cycle is not unlike the shuffling gait of a terrestrial bipedal vertebrate. The elevator and protractor muscles of the forward-moving half exert their thrust against the static reaction between the other half of the foot and the ground. The forward thrust is equal and opposite to the backward thrust applied to the anchor. Although the pedal musculature is extremely complex, there can be little doubt that the muscles transmit their forces by means of the hydraulic system of the pedal haemocoel.

Retrograde waves: pedal anatomy

There are few detailed descriptions of the anatomy of the feet of gastropods in which structure is related to function. Since locomotion is discussed here first in relation to *Patella*, it is convenient to consider briefly its pedal anatomy. In *Patella* most of the muscle fibres of the foot originate on the shell and insert laterally or

Fig. 2.9 Longitudinal section through the foot of the limpet, *Patella*, showing epithelium of sole, spherical haemocoelic cavities and dense muscular layer dorsally (stipple). Principal muscle fibres are arranged transversely and dorso-ventrally.

ventrally on the basement membrane of the dermal epithelium.[59] There are four groups of muscle fibres (Fig. 2.9): (a) those passing directly to the sole on the same side as their origin, called dorso-ventral muscles; (b) those going to the sole and lateral margin of the foot on the side opposite to their origin, called transverse muscles; (c) muscles running longitudinally from their origin to the snout or pedal epithelium; (d) spiral muscles which pass in a spiral course from shell to sole. The bulk of the foot consists of the dorso-ventral (70% of the musculature) and transverse muscles (25%). Very few muscle fibres run longitudinally in *Patella*; some pass from the anterior end of the shell to be inserted near the posterior margin of the foot, but there are none situated immediately above the sole. That longitudinal muscles are not required in locomotion has been demonstrated by making a series of transverse incisions across the foot. These did not affect the passage of pedal locomotory waves. An important component of the foot is the blood in its haemocoel. Blood flows from the cephalic sinus into two pedal sinuses, whence it diffuses into the pedal haemocoel. This is a region, extending some distance above the sole, containing very distinct spherical spaces each of about 10 μm diameter (Fig. 2.10). Very little loss of blood

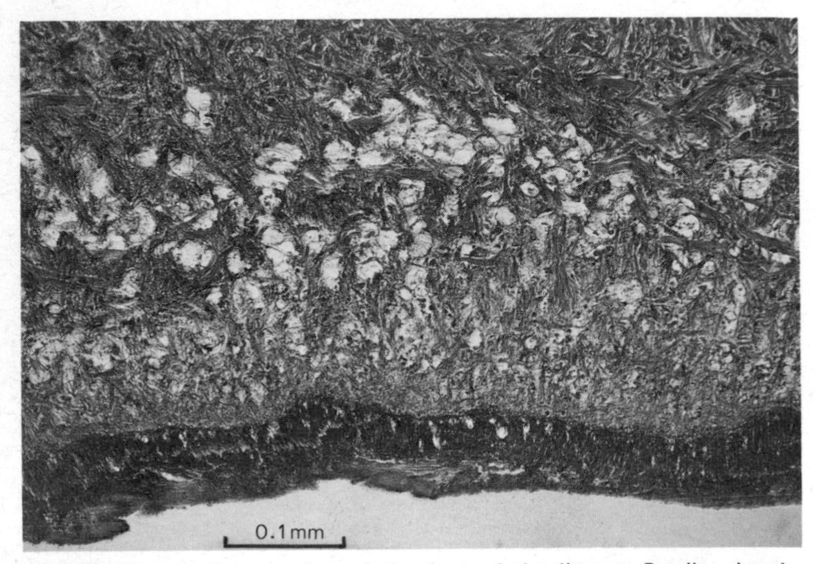

Fig. 2.10 Longitudinal section of the foot of the limpet, *Patella*, showing epithelium of the sole and muscular region above with numerous spherical vesicles. Note absence of longitudinal muscle near sole. (photograph Dr H. D. Jones)

occurs in respect of the transverse incisions referred to above. How exactly this is achieved is not clear, but contraction of adjacent muscles may possibly suffice to seal the wound, since the haemocoel consists of a large number of small spaces. The importance of the blood should not be underestimated, for it forms virtually the only skeletal support of the foot. Molluscs characteristically possess a shell, an exoskeleton which partially surrounds the tissues, but it is only relatively rarely employed for muscular antagonism without blood in the body or water in the mantle cavity participating as a fluid skeleton.

Mucous glands are a prominent feature of the feet of all gastropods and mucus is undoubtedly secreted in copious amounts in association with muscular locomotory waves.

Retrograde waves: locomotory mechanics

When limpets are allowed to settle and crawl on glass plates, alternating sets of waves may be observed as pale areas of the sole to pass backwards on either side of the foot during forward locomotion (ditaxic retrograde waves). These waves, which are produced by the contraction of the dorso-ventral muscles, lift the sole off the substrate and stretch the epithelium, as is indicated in diagrammatic section (Fig. 2.11). The sole shortens to its original length as it is

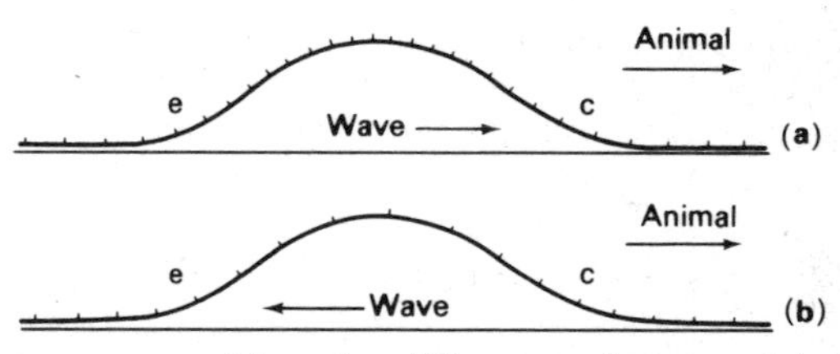

Fig. 2.11 Diagram summarizing the differences between (**a**) direct and (**b**) retrograde pedal waves. The amount of compression (c) or extension (e) of the sole of the foot is indicated by markers. The leading edge of a direct wave is associated with compression, that of a retrograde wave with extension. The lagging edges show the reverse of this. (after Lissmann[65])

lowered at the end of the step. Analysis of cine film in which the foot was marked by shallow incisions has allowed these movements to be demonstrated (Fig. 2.12a). The extension of the region which is off the ground, and the anchorage of parts of the foot which are at their shortest length, is exactly parallel to the mechanism described for retrograde locomotion in worms. Retrograde locomotion is

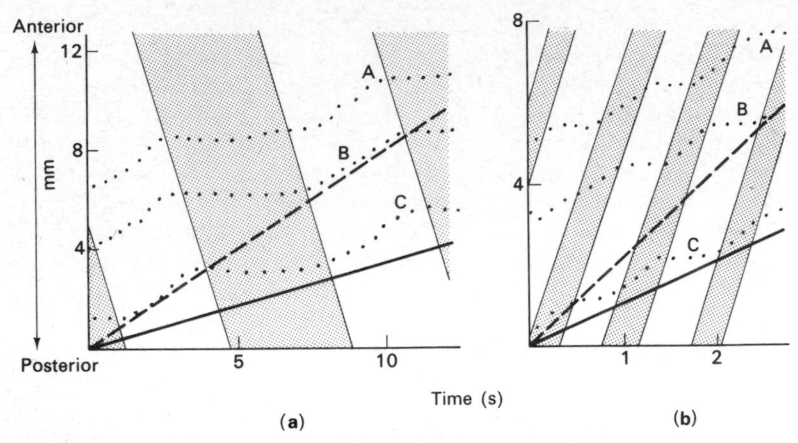

Fig. 2.12 Graphs obtained from cine film of the forward locomotion of (**a**) *Patella*, (**b**) *Helix*. Each shows the forward motion (dotted) of 3 points (A–C) arranged antero-posteriorly on the foot and relative compression of the sole by stipple. Compression occurs with a retrograde wave (a) when the sole is attached to the substrate and with a direct wave (b) when moving forwards. Heavy line represents average speed of the animal and the broken line the average speed of a point in forward motion (a, from Jones and Trueman;[59] b, from Lissmann[65])

always characterized by the shortening of the region of the propellor surface applied to the substrate.

Two levers, arranged as in figure 2.13 and coupled by transducers to a pen recorder, enabled movements of the sole to be recorded accurately during locomotion, anterior motion always corresponding to the raising and extension of the sole. It proved possible to calibrate the vertical displacements and to obtain a mean value of 0.02 cm for the amount that the foot is raised during the passage of a wave. Photographic analysis shows that at any instant during locomotion, about 58% of the foot of *Patella* is raised off the substrate and is moving forwards. If the area of movement is multiplied by the depth of a wave, then in an adult limpet of pedal area 4 cm^2 there is an upward displacement of about 0.046 ml which is probably represented by displacement of blood from the foot. The blood volume of a limpet with a foot of this size is approximately 3 ml so that the amount displaced is only a very small proportion of the whole and is unlikely to cause much variation in the pressure in the haemocoel.

It was not possible to record pressure from within the pedal haemocoel because of the small size of the blood spaces, but it is reasonable to assume that as the foot is raised by contraction of the

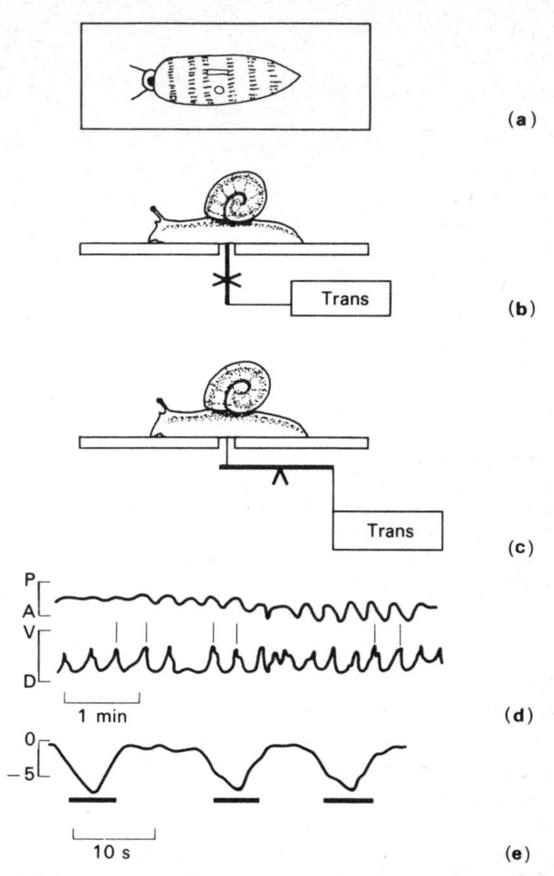

Fig. 2.13 (**a–c**) Arrangement of levers for the simultaneous recording of horizontal and vertical displacements of the snail's foot during locomotion (after Lissmann[65]). (**d**) Record from the sole of *Patella*, horizontal movement as posterior (P) or anterior (A) displacement, vertical movement as ventral (V) or dorsal (D) displacement. Vertical bars are visual observations of compression of the sole coincident with lever. (**e**) Pressure recording ($\times 10^{-2}$ N cm^2) from beneath the foot of *Patella* during forwards locomotion; horizontal bars indicate passage of waves (elevation of sole) over pressure cannula. (from Jones and Trueman[59])

dorso-ventral muscles, each part of the haemocoel is somewhat compressed against the musculature above and becomes deformed (Fig. 2.14). Increase in the lateral dimensions of the foot is prevented by the transverse muscles and, provided there is no great loss of blood, as is suggested by experiments involving cutting the foot, then deformation of spherical haemocoel cavities must result in

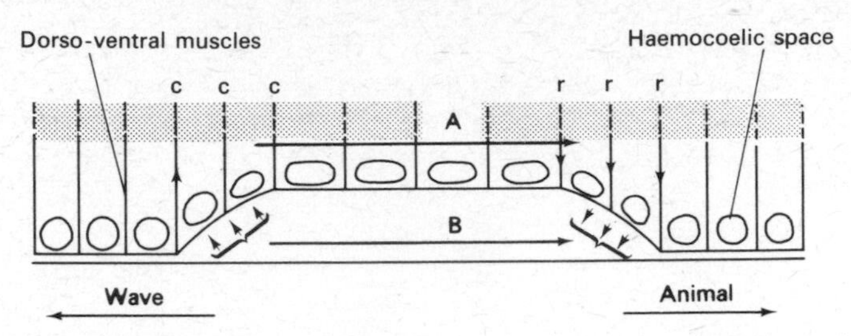

Fig. 2.14 Parasagittal section of the foot of *Patella* showing factors involved in progression of a retrograde locomotory wave. Haemocoel spaces are distorted and the sole is drawn up by contraction (c) of the dorso-ventral muscles. The suction thus generated beneath the sole (B), together with the pressure in the haemocoel (A), antagonizes the dorso-ventral muscles when they relax (r) and draws the sole of the foot down. The stippled region represents a thick layer of muscle fibres. (after Trueman[118])

elongation of the epithelial surface. Relaxation of the dorso-ventral muscles at the trailing edge of a pedal wave (Fig. 2.14, r) results in the lowering of the sole with the restoration of the haemocoel cavity to a spherical shape and the epithelium to a more compressed condition. The transference of force from the leading to lagging edge of the wave via the haemocoel is indicated by arrow A. A small standing blood pressure in the pedal haemocoel antagonizes the dorso-ventral muscles, and, when these relax, presses the sole to the substrate.[118]

If a limpet crawls over a small hole in a perspex plate, through which a cannula is connected to a pressure transducer, negative pressures are recorded as the dorso-ventral muscles contract to raise the foot off the substrate (Fig. 2.13e). Thus the lifting of the sole of the foot associated with the passage of each locomotory wave produces a suction-like effect on the mucus beneath. The dorso-ventral muscles contract at the leading edge of the pedal wave (Fig. 2.14c) and, by means of the negative pressure produced (arrow B), antagonize the dorso-ventral muscles at the lagging edge, so that as these relax (r) the sole is drawn down onto the substrate. Progression of the pedal wave in *Patella* is thus brought about by the successive contraction and relaxation of the dorso-ventral muscles. The antagonism between the muscles at the leading and lagging edges of the wave involves both the internal (haemocoelic) (A) and external (B) hydraulic systems; the presence of longitudinal muscle fibres adjacent to the sole is not required, except possibly at the trailing edge of the foot. Some of the work done by the contracting dorso-ventral muscles is thus expended in the restitution of other fibres to their resting length, but most of the remaining work is converted into a propulsive force by the hydraulic properties of the foot. As already stressed, lateral extension of the foot must be strictly controlled. Numerous transverse muscle fibres perform this function in *Patella*; if they are cut experimentally, the speed of locomotion is greatly reduced.

Another problem concerning the limpet's locomotion is how the anterior part of the foot is advanced at the start of a retrograde wave. The solution is found by a continuation of the preceding paragraphs. If the pedal haemocoel at the front of the foot is vertically narrowed with lateral extension restricted by transverse muscles, and blood is not allowed to escape, then, with the posterior part of the foot anchored, forward extension must follow.

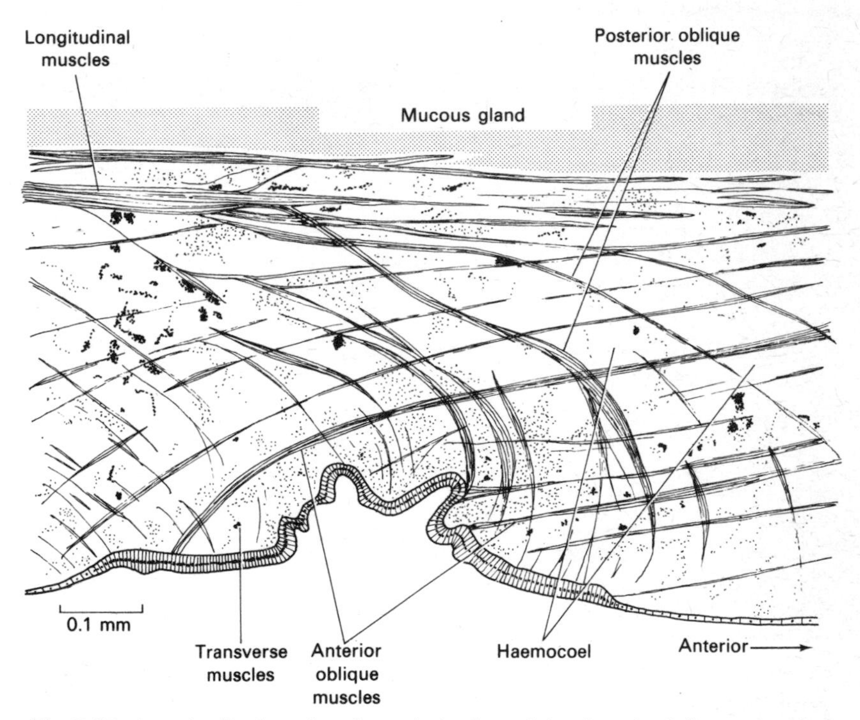

Fig. 2.15 Longitudinal section through the foot of the slug, *Agriolimax*, showing the passage of a direct pedal locomotory wave. The sole and pedal haemocoel above is compressed in the region of the wave. Immediately beneath the anterior mucous gland (stipple) lie longitudinal muscles from which anterior and posterior oblique muscles pass to the sole. Numerous transverse muscles are also shown passing through the section.

Direct waves: pedal anatomy

The anatomy of the foot of pulmonates is comparable to that of *Patella* in respect of the epithelial sole, haemocoelic cavities and the presence of transverse muscle fibres. There are, however, no muscles orientated dorso-ventrally but there is a substantial layer of longitudinal muscle from which fibres pass down obliquely from anterior and posterior to the sole (Fig. 2.15). The section in Fig. 2.16, which shows two direct locomotory waves with the epithelium longitudinally compressed, was obtained by dropping a crawling slug into liquid nitrogen to achieve almost instantaneous fixation.[57,58]

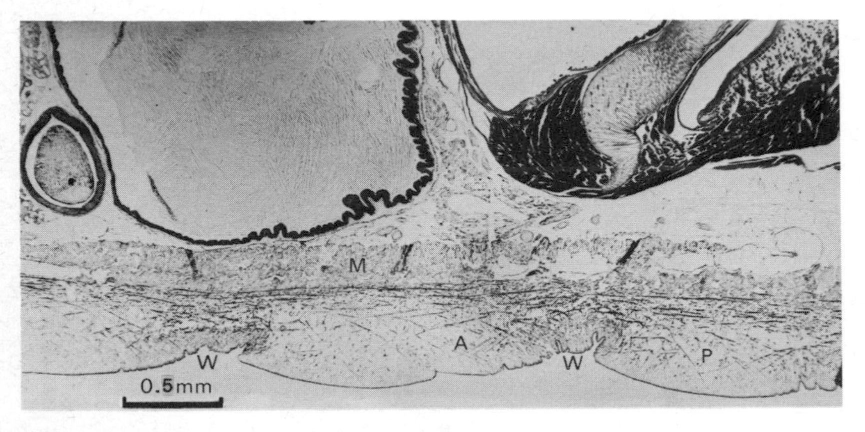

Fig. 2.16 Longitudinal section of the foot of the slug, *Agriolimax*, showing two direct locomotory waves (w) in which the sole is constricted longitudinally when raised from the substrate, anterior (A) and posterior (P) oblique muscle fibres with mucous gland (M) and viscera above. (photograph Dr H. D. Jones)

Mucus is important in the locomotion of all snails and Runham and Hunter[93] have described the secretion of two types in the slug, *Limax*. The first, from the anterior pedal mucous gland (situated above the longitudinal muscle layer), is of high viscosity and forms a continuous film beneath the foot; the second, of more fluid consistency, is secreted by glands opening over the whole surface of the sole. The latter type provides the fluid medium necessary for the functioning of locomotory waves, while the film of viscous mucus ensures a secure anchorage for the static regions of the sole.

Direct waves: locomotory mechanics

When the snail *Helix* starts to move, a dark transverse band appears across the foot anteriorly and passes forward. Other dark bands arise posteriorly and the sole of the foot soon displays a pattern of a series of moving bands. These move in the same direction as the foot and are thus direct locomotory waves. Lissmann[65,66] (summarized by Gray[43]), using the same technique as illustrated in Fig. 2.13, demonstrated that the dark bands represent longitudinally contracted regions of the foot and that these are raised from the substrate. This is the converse of what has been described in respect of retrograde waves, where the dark bands are the regions of the epithelium that are attached to the substrate. With direct waves, the foot is anchored to the substrate at maximum extension of the sole (Fig. 2.11a) in a manner directly comparable to

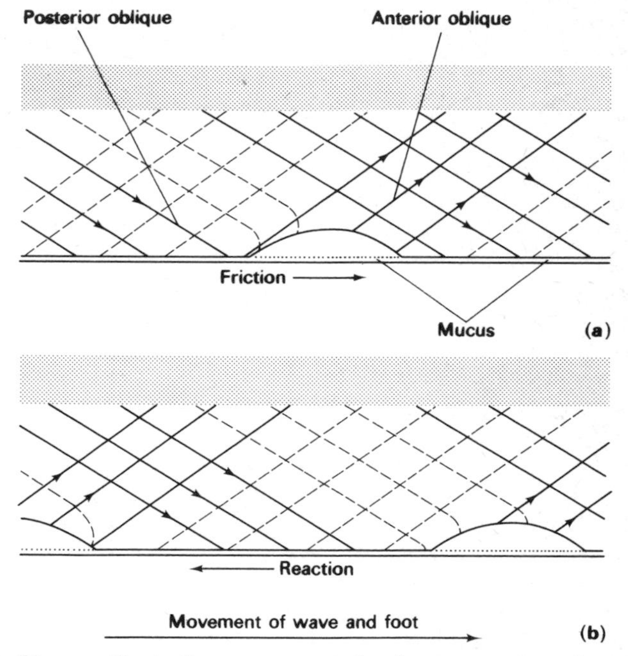

Fig. 2.17 Diagram illustrating movement of a direct wave along the sole of the foot of a slug. The layer of longitudinal muscle, indicated by stipple, gives anterior and posterior oblique muscle fibres to the sole. Broken line indicates regions of these muscles where relaxed, solid line and arrowheads where contracted. A thin layer of mucus is shown to be continuous beneath the foot and increases in volume beneath the pedal wave.

that described above (p. 25) for the locomotion of worms. The nature of the pedal haemocoel allows the foot to adhere to the substrate when fully extended. By analysis of cine film, Lissmann has further demonstrated the forward motion of three marked points in the foot of *Helix pomatia* (Fig. 2.12b) where the sole is compressed. Examination of stained longitudinal sections of the feet of slugs (Fig. 2.16) confirms these observations and allows some analysis of the functioning of muscles and of the fluid skeleton.[57]

The progress of a snail depends on the animal applying sufficient force to the ground in a posterior direction (static reaction) to overcome the sliding friction of parts moving forward. The foot only moves forwards when raised from the substrate, probably by tension in the anterior-oblique muscles (Fig. 2.17), the cavity beneath the sole containing fluid mucus which allows movement. Between waves the extended part of the sole is anchored to the ground by a layer of more viscous mucus; contraction of the posterior oblique muscles inserted into this part of the sole would then set up static reaction forces and draw the body forward. Thus a forward frictional force followed by backward thrust would be associated with each pedal wave, depending on whether the sole is respectively detached from or attached to the substrate. This was tested by allowing a snail to crawl over a bridge of a kind similar to that originally designed by Lissmann[66] but using a transducer to record force (Fig. 2.18).

The general picture obtained in recordings is first of sliding friction as the front of the head is pushed onto the bridge, second a series of oscillations representing friction and reaction associated with each pedal wave. Using a broad bridge (b), the static reaction reaches a maximum with the anterior part of the snail over the bridge whilst the weight of the shell and viscera are being pulled forward. During the passage of the tail, which is effectively being pulled along behind, sliding friction is at a maximum. But when a narrow bridge, of about the same dimensions as a pedal wave, is employed, these gross effects are removed, and the oscillations are approximately about the zero line. The forces involved again diminish as the tail is pulled over the bridge although the frictional component remains high. The large deflection (in c) indicates increase in the force exerted by the posterior oblique muscles as the shell moves laterally.

To summarize this argument, it may be considered that regions of the sole at rest, forming anchorages, elicit the static reaction from the ground, while regions of the foot which are moving forward elicit a sliding frictional resistance. During movement at constant

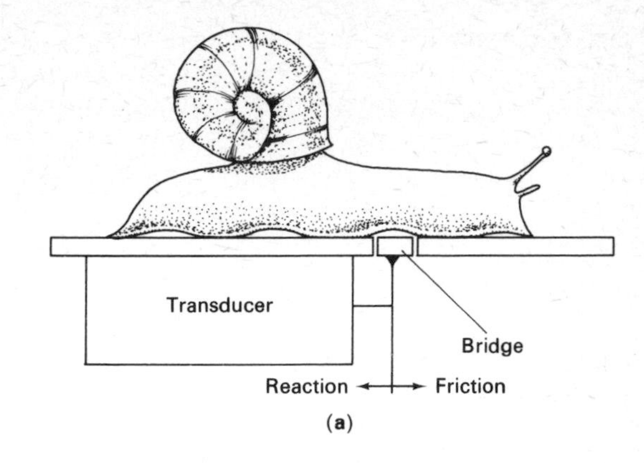

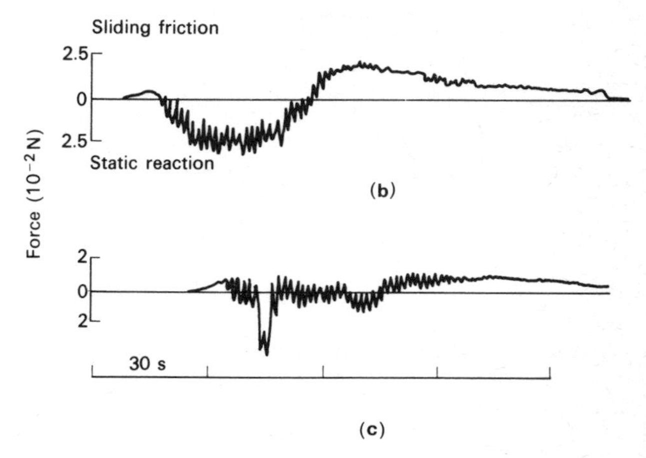

Fig. 2.18 (**a**) Arrangement for recording forces developed in locomotion between foot of *Helix pomatia* and substrate. (**b**) Recording obtained using wide bridge (about 1 cm); sliding friction occurs when bridge is pushed to the right, i.e. forwards, static reaction when pulled to the left, i.e. backwards. (**c**) Similar recording of forces using narrow bridge comparable in size to pedal waves; oscillations represent each wave, major deflection due to shell moving from one side of the snail to the other.

speed, the static reactions from the *points d'appui* must be equal but opposite to the dynamic resistances of the parts of the body in motion. This process is facilitated by the secretion of viscous and fluid mucus for the purpose of adhesion and movement respectively.

Other locomotory waves

When creeping slowly forward *Helix* leaves behind a continuous trail of mucus, for most of its foot is in contact with the ground; but some members of this genus adopt a different gait when moving at speed and leave a discontinuous trail.[43] The fast loping gait involves large retrograde waves which occur simultaneously with the normal, much slower and smaller, direct waves (Fig. 2.19). The sole is

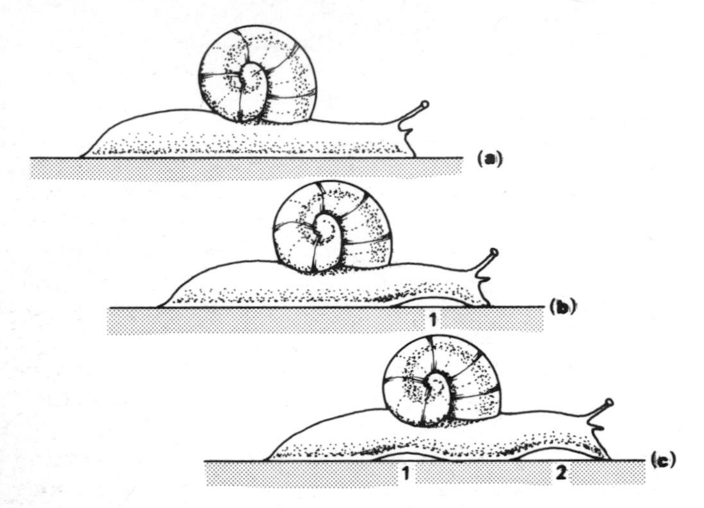

Fig. 2.19 Fast loping movement of snail, (**a**) normal direct locomotory waves; (**b**) and (**c**) adoption of fast loping gait with development of large retrograde waves. (after Gray[43])

attached to the substrate at its shortest length and greatest width, as is to be expected with a retrograde wave. Large locomotory waves of this kind have also been observed during burrowing in sand by *Polinices* and *Natica*,[115] where the posterior part of the foot (mesopodium) acts first as an anchor for pedal extension and is subsequently drawn forward when the anterior region of the foot or propodium is anchored (Fig. 2.20 and 3.19). Perhaps the logical conclusion to this type of locomotion is that adopted by *Strombus gigas*, the West Indian giant queen conch, which rests its shell on the ground as an anchor during pedal extension and subsequently lifts the shell and draws it forward while the foot adheres to the substrate.

The use of anterior and posterior regions of the foot of gastropods as alternate anchors during locomotion resembles the

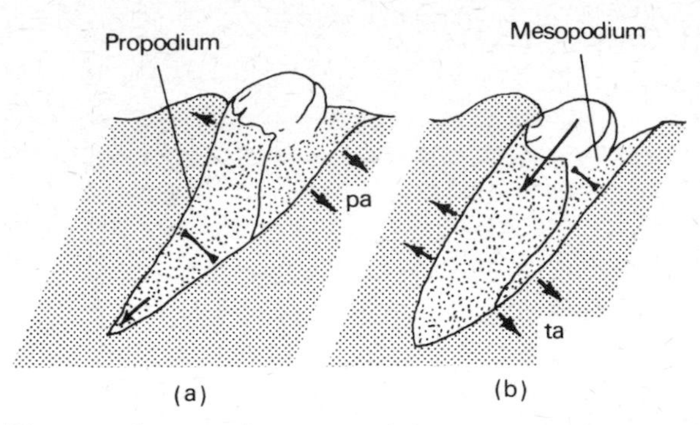

Fig. 2.20 Diagram of successive stages of the burrowing of *Polinices* into sand, (stipple). (**a**) Mesopodium distended forming penetration anchor (pa) which allows protopodium to extend into the sand by contraction of dorso-ventral muscles. (**b**) Propodium dilated, by contraction of mesopodium, to form terminal anchor (ta) allowing shell and mesopodium to be pulled into sand. (after Trueman[115])

looping movements of leeches. In the latter anchorage is by suckers, in the gastropods by mucous adhesion. The major difference between pedal waves passing along the body and a galloping or loping means of progression is the number of waves of contraction in relation to body length. In galloping the ratio approaches 1: 11, whereas it may be as high as 10: 1 in respect of pedal waves. In essence, all methods of muscular locomotion developed by soft bodied animals over hard surfaces demand stability of anchorage of part of the body against forces of sliding friction developed in other regions. This anchorage is successfully achieved as a result of the thickening of the animal, or of the pedal surface, by longitudinal shortening and the production of retrograde waves. Both on the grounds of phylogenetic distribution and in respect of the simple mechanical principles involved, retrograde waves would appear to be more primitive than direct locomotory waves. Locomotion by retrograde waves is the consequence of an animal raising the anterior end of the body or foot, and extending it forwards. By contrast, locomotion by direct waves initially requires the tail to be raised and pulled towards the head as a preliminary to a locomotory wave moving forwards along the body.

In the locomotion of both *Strombus* and *Pomatias* (p. 28) the use of pedal locomotory waves has been greatly modified from what occurs in the flatworms and nemertines and clearly involves the use of a well developed haemocoel. This, indeed, is the essential underlying feature which has ensured the success of pedal locomotion by gastropods.

3

The Penetration of the Substrate by Soft-Bodied Animals

INTRODUCTION

Movement of soft-bodied animals over substrates has been shown in Chapter 2 to be brought about either by ciliary movement or by the anchorage of one part of the body while another region is drawn or thrust forward by muscular contraction. The forces involved are small and anchorage by mucoid adhesion is generally sufficient for progress to be made. Increase in the power used in locomotion may be brought about by increasing the area of the propellor surface so that a flattened shape is an advantage. Circular muscle fibres can play little part in locomotion with such a shape for their contraction must tend to make the animal more nearly circular in cross section. This has the effect of reducing the area of contact with the substrate, hence diminishing the strength of anchorage and, accordingly, reducing the force available for thrusting forward. A circular cross-section is, however, ideally suited for movement into or through a substrate, since the body wall is in contact with the substrate on all sides and all muscles may contribute to the locomotory force without loss of anchorage. Many soft-bodied burrowing animals are circular in cross section, for example, the anemone *Peachia.* the lugworm *Arenicola,* and the bivalve mollusc *Ensis,* and the possession of a true fluid skeleton in each of these enables them to make powerful movements. The forces developed by muscular contraction are commonly remote from the point of their application, and are transferred by the body cavity being used as a hydraulic system. The body cavity used differs according to phylum. The Coelenterata utilize the coelenteron as in *Peachia*, the Annelida use the coelom as in *Arenicola*, and the Mollusca employ the haemocoel as in *Ensis*.

These different kinds of body cavities all function as hydraulic organs in a similar manner and exhibit their maximal development in respect of the burrowing habit.

MECHANICAL PRINCIPLES OF BURROWING

Animals that burrow may be placed in two broad categories, those whose bodies are completely hard, for example crabs or heart urchins, and those that are soft-bodied such as a worm, or utilize soft processes of the body in burrowing as do clams. Apart from those

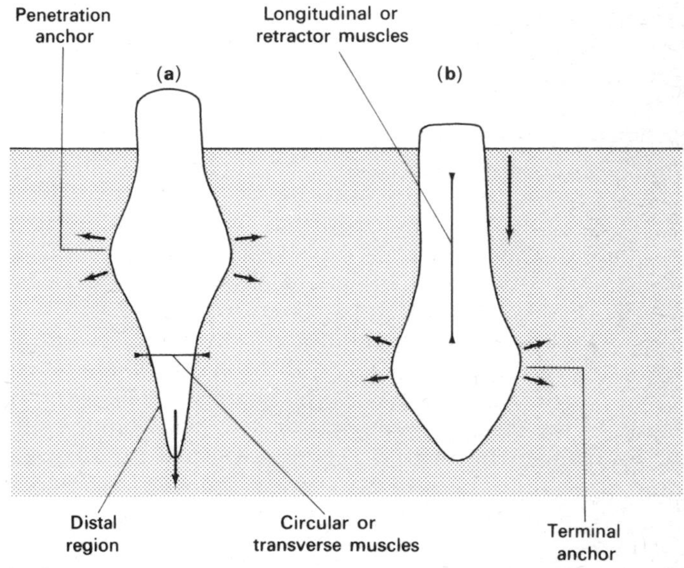

Fig. 3.1 Diagrams showing the two principal stages in the burrowing process of a generalized soft-bodied animal. (**a**) Formation of penetration anchor which holds the animal when the distal region is elongated by contraction of the circular or transverse muscles. (**b**) Dilation of distal region producing a terminal anchor which allows contraction of longitudinal or retractor muscles to pull the animal into the substratum. Maximum pressures are developed in the fluid system at this stage. (from Trueman and Ansell[121])

worms and chordates which use undulatory swimming movements for burrowing, for example amphioxus,[129] all soft-bodied animals burrow by means of an essentially similar mechanism. This is based on the formation of two types of anchors, first the 'penetration anchor' by dilation of an upper part of the body and secondly the 'terminal anchor' by distension of the extremity of the body (Fig. 3.1). The tip of the animal is forced downward by contraction of circular or transverse muscles and at this instant the penetration anchor grips the burrow, so preventing the animal from being thrust upwards. The stronger the penetration anchorage the greater is the force that may be brought to bear against the substrate. With weak anchorage only small forces can be employed to extend the burrow if the animal is not to be pushed backwards. The terminal anchor is formed by the body wall being pressed outwards against the substrate so as to obtain sufficient adhesion to allow the upper part of the body to be drawn into the burrow by the contraction of longitudinal or retractor muscles. These two types of anchors are applied alternately until burial is complete so that progress generally appears to take the form of a series of steps.

Many worms make burrows in sand of a semi-permanent nature, in which the walls may be consolidated by mucus. They may crawl along these burrows by locomotory movements not unlike those of the burrowing process. One part of the body may be dilated, with contraction of longitudinal muscles forming *points d'appui*, equivalent to the penetration anchor, whilst anteriorly the circular muscles contract to cause extension, as for example in the earthworm. These contractions progress along the body as a retrograde peristaltic wave from head to tail in all septate worms, e.g. the earthworm (p. 60). Indeed, most burrowing animals of circular cross section utilize peristaltic locomotory waves but in some, notably *Arenicola* (p. 56), *Peachia* (p. 52) and *Polyphysia*, the waves are direct and passs along the body to the head or foremost extremity. Discussion of the displacement of a worm relative to the substrate (p. 26) has shown that with a retrograde wave the segments are anchored when fully shortened, i.e. are at their shortest and fattest, but that with a direct wave the segments must be anchored when fully extended. In a septate worm, such as an earthworm, each segment would operate at constant volume, extension being accompanied by reduction of diameter, a condition unsuited to obtaining anchorage in a burrow. Thus for an earthworm retrograde waves are the only possibility for producing forward motion in a burrow. The conditions are somewhat different in respect of a worm without trunk septa, for example the polychaete, *Polyphysia*, or in an animal with a continuous body cavity such as the anemone *Peachia*, for the longitudinal and circular muscles are no longer segmentally arranged about a small constant volume of fluid. The entire body fluid is effected by all contractions of these muscles and maximum extension of part of the body may be achieved at the same time as maximum girth.

Polyphysia lives in flocculent mud and moves along its burrow by means of direct waves. These take the form of peristaltic waves involving the simultaneous contraction of circular and longitudinal muscles of the body wall. By measuring the dimensions of the worm during locomotion, Elder[34] has shown that the length of each segment is minimal as the constriction passes, and that behind the wave the length of each segment increases simultaneously with the circumference of the body wall. In addition to the musculature, the body wall of *Polyphysia* contains a well developed connective tissue layer consisting of a three-dimensional collagen lattice and radial elastic fibres (Fig. 3.2).[33] Deformation of the lattice occurs during

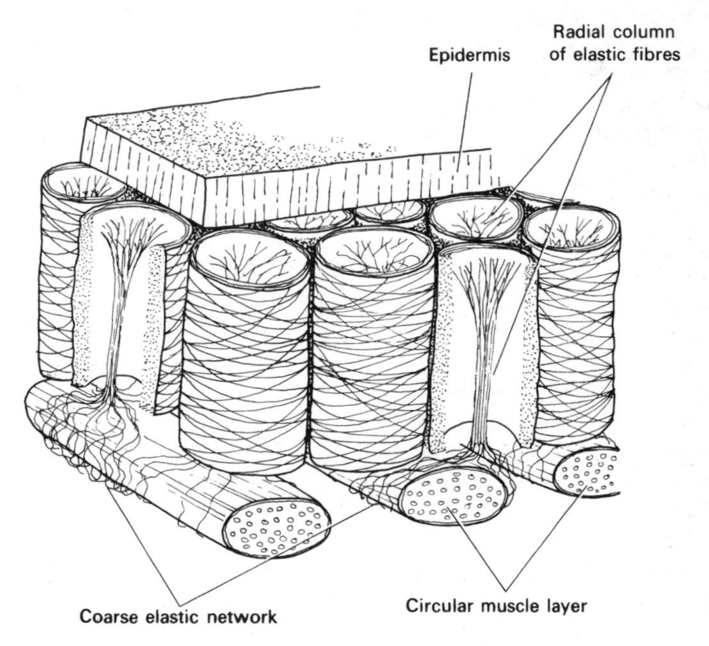

Fig. 3.2 Diagram illustrating the fibrous construction of the connective tissue layer in the body wall of *Polyphysia*. Radially orientated units of the three dimensional collagen lattice are shown extending between the circular muscle layer and the epidermis. Radial columns of elastic fibres extend from the coarse elastic network around the muscle bundles, to the epithelial basement membrane. (from Elder[34])

peristalsis, while the elastic fibres oppose radial distension, control the folding up of the epidermis and are thought to facilitate the return of the body-wall after the passage of each constriction. Elder has shown that most, if not all, of the coelomic fluid is displaced during the passage of a peristaltic wave to contribute to the dilation of adjacent regions. The muscular contractions of peristalsis generate a pressure in the coelomic fluid which is deployed in the adjacent segments of the body, there causing extension of both circular and longitudinal muscles. Thus segments are dilated and lengthened simultaneously so as to be anchored against the burrow walls when maximally extended. *Polyphysia*, accordingly, fulfils the conditions of anchorage for forward locomotion by means of direct waves. This

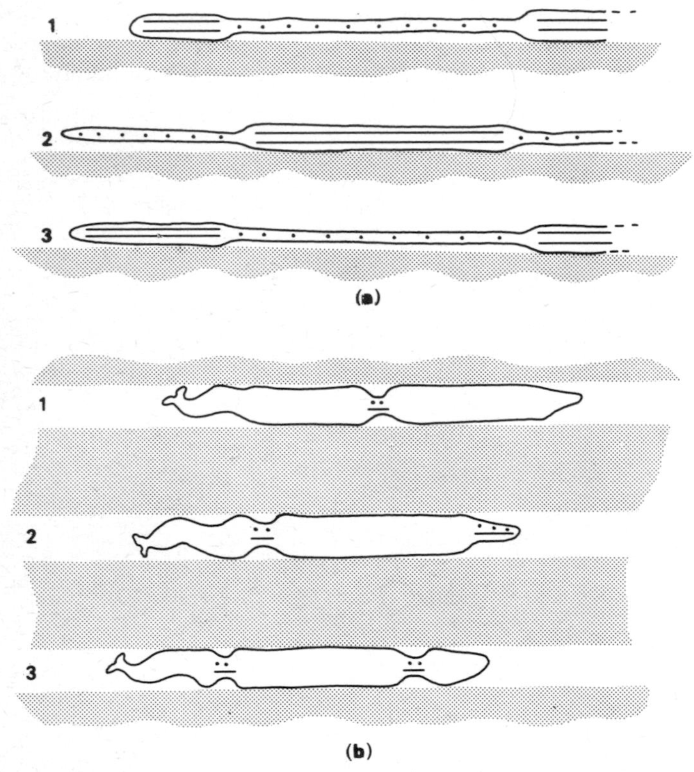

Fig. 3.3 Diagram comparing aspects of the locomotion to the left of (**a**) *Lumbricus*, by retrograde waves; (**b**) *Polyphysia*, by direct waves. The muscle layers, longitudinal or circular, are indicated where they are predominantly contracted. In *Lumbricus* the *points d'appui* are in regions of longitudinal muscle contraction, in *Polyphysia* anchorage is obtained in regions of relative relaxation of both circular and longitudinal muscles. (from Elder[34])

may be contrasted with retrograde locomotion in the earthworm, where the septate condition precludes anchorage of each segment except at maximum girth when the longitudinal muscles are fully contracted (Fig. 3.3). Locomotion in a burrow by means of direct waves is only possible in a worm with a continuous trunk coelom, for this allows parts of the body wall to achieve maximum girth and maximum segmental length at the same instant.

BURROWING ACTIVITY

Most animals that burrow progress into the ground in a step-like series of movements. The events associated with each step are repeated cyclically and are referred to as a 'digging cycle', a term first used in respect of bivalve molluscs. A series of cycles is well illustrated from recordings of *Donax* (Fig. 3.4.a), made with a thread

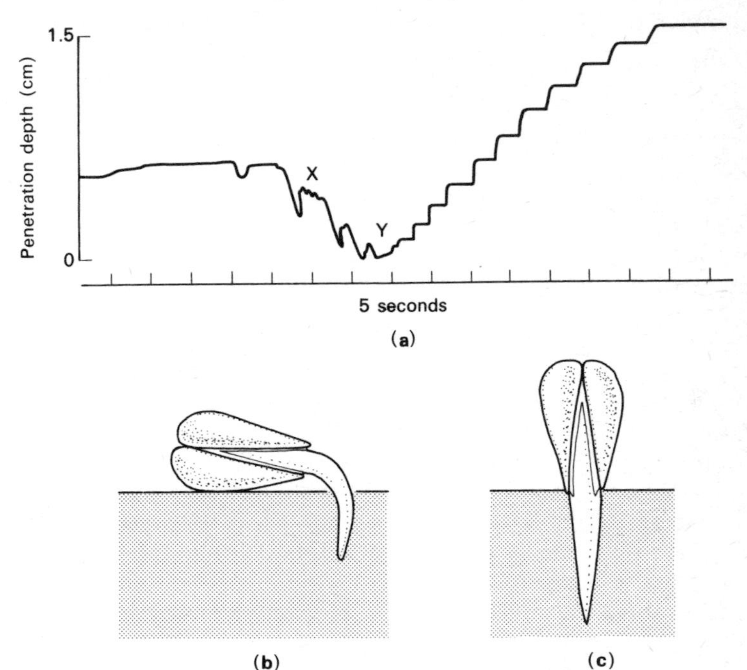

Fig. 3.4 (**a**) Recording of a complete digging period of the bivalve *Donax vittatus* recorded by attaching a thread from the posterior of the shell to an isotonic transducer. Two phases are shown, namely, probing of the foot (x) to penetrate sand with animal lying on side (b), followed at Y by sequence of digging cycles, upstrokes represent penetration of shell into sand (cm). At Y the foot is sufficiently buried to pull shell into the erect position shown in (**c**).

attached from the posterior of the shell to an isotonic transducer. Similarly, the term 'digging period' was also originally used in respect of bivalve molluscs, members of the Veneridae,[4] and is now applied to the duration of burrowing activity in all animals, from the beginning until a stable position is attained in the substrate. The digging period, which comprises a number of digging cycles, may be conveniently divided into two parts: (i) initial penetration, when cycles occur only sporadically; (ii) movement into the substrate, when the cycles follow in regular succession.

INITIAL PENETRATION OF SUBSTRATE

Donax

The division of the digging period into two phases may first be usefully considered in respect of the bivalve *Donax* (Fig. 3.4). The digging period commences with the clam lying on its side (b), the foot probing to penetrate the sand. During this first part of the digging period the force with which the foot can enter the substrate is limited since the only resistance to the backthrust derived from probing, i.e. the penetration anchor, is due to the weight of the shell on the sand. Eventually, at Y in the recording after several failures, the foot has penetrated sufficiently to be able to exert on the sand a lateral force which is sustained as a sufficiently strong anchor to allow the shell to be drawn into an upright position (c) and held there. This marks the beginning of the second phase of the digging period, when cycles follow in rapid succession.

During the first part of all digging periods, before the animal has entered the burrow, the penetration anchor cannot be applied in the manner which occurs subsequently. Only weak penetrative scraping movements may be made by animals which are circular in cross-section with little contact, and therefore adhesion, to the surface of the substrate. Those without dense hard parts, such as anemones or polychaete worms, have little weight in water and exhibit various adaptations to facilitate initial penetration. With greater weight (for example, a molluscan shell), or with multiple appendages (for example, the mole crab, *Emerita*[119]), larger forces may be brought to bear and specialized scraping devices have less importance.

Arenicola

When a lugworm is placed on the surface of sand it invariably begins immediately to burrow by means of repeated proboscis eversion, the mouth being turned down onto the surface of the sand. The proboscis consists of the buccal mass bearing short tooth-like papillae, and a papillate pharynx (Figs. 3.5 and 6). This structure and the oesophagus are suspended from the body wall by a strong sheath of retractor muscles and a thin gular membrane. These represent the first intersegmental septum. Immediately behind, two septa divide the trunk coelom somewhat imperfectly, for they are

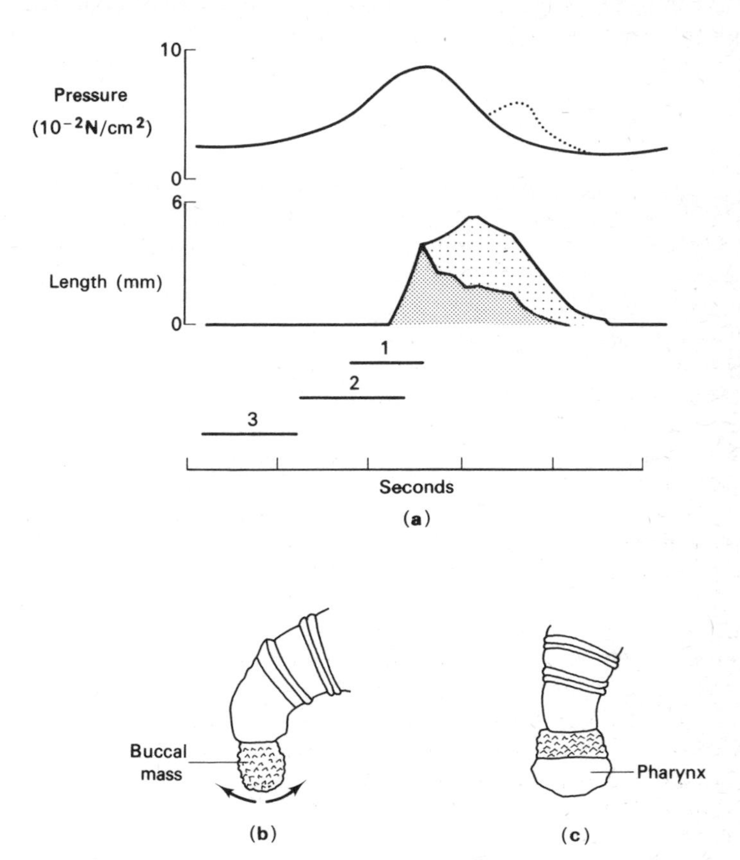

Fig. 3.5 Movement of proboscis and flange activity in *Arenicola*. (**a**) Pressure ($\times 10^{-2}$ N cm^2) recorded from general trunk coelom, dotted line represents extra pressure pulse from head coelom on retraction of buccal mass; length (mm) of buccal mass (stipple) and pharynx (coarse stipple) extruded; horizontal bars represent keeled appearance of annular flange of segments 1, 2 and 3 respectively. (**b**) eversion of buccal mass with centrifugal movement, note keeled appearance of anterior flange. (**c**) Complete eversion of proboscis with pharynx and buccal mass visible. (after Seymour[100])

thin and perforated and cannot impede transfer of fluid or pressure. The coelomic fluid of the trunk of *Arenicola* is enclosed within the tubular body wall, with outer circular and inner longitudinal muscle fibres and has no other septal divisions; thus the whole trunk coelom acts as a single hydraulic organ (Fig. 1.2). The buccal mass emerges from the head at regular intervals and pushes the sand aside with a centrifugal movement (Fig. 3.5b arrows) which is essentially a scraping action. This is then followed by eversion of the pharynx. The action of the proboscis serves to penetrate, thrust aside and drag back the sand and to draw the head into the burrow so formed. Only very occasionally has the proboscis been observed to swallow sand during burrowing, and it must in no sense be thought of as having the action of a cork borer.

Chapman and Newell[24] calculated that the maximum thrust of a worm weighing 2.5 g on the surface of the sand is equal to about 5.2×10^{-3} N and that, while the proboscis pressure can hardly

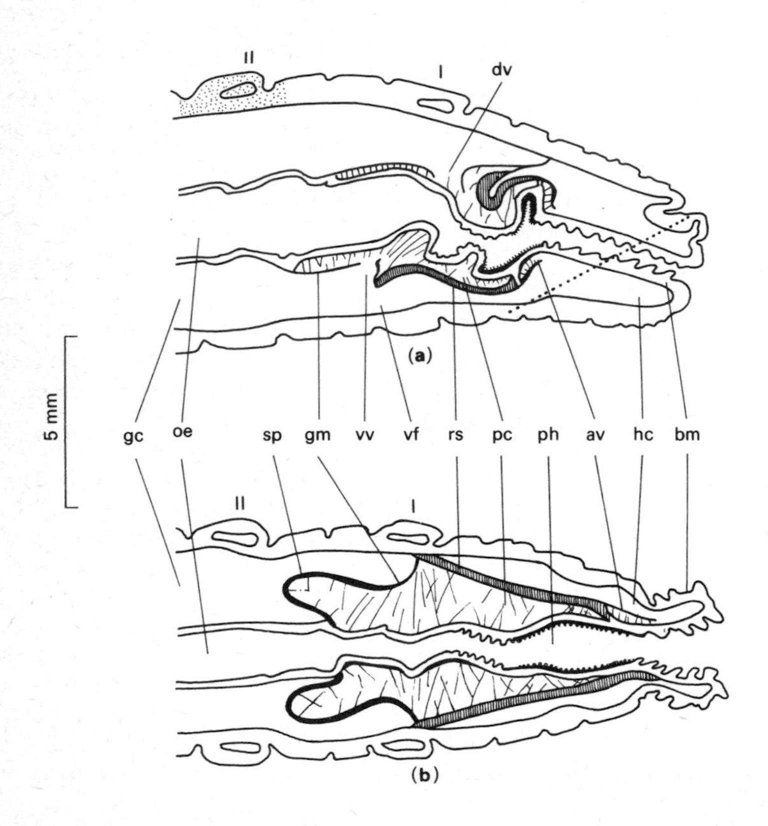

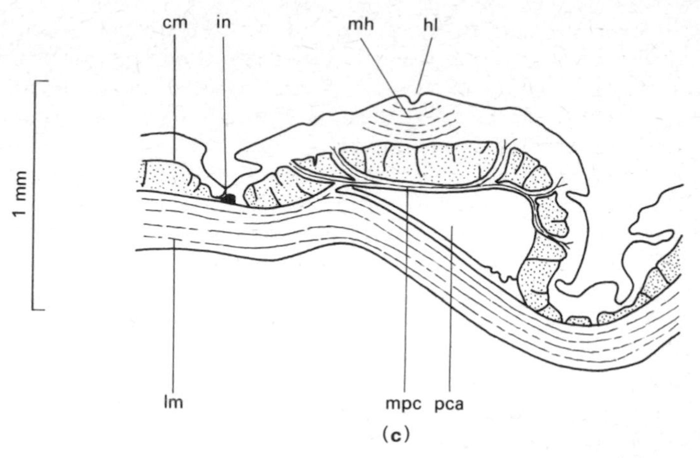

Fig. 3.6 Diagrammatic sections of the anterior end of *Arenicola marina*, (**a**) median sagittal section; (**b**) horizontal section; (**c**) nearly median sagittal section of the second chaetigerous annulus (area shown stippled in (a)). In (a), the proboscis is withdrawn; in (b) the buccal mass is everted. I, first chaetigerous annulus; II, second chaetigerous annulus; av, anterior valve; bm, buccal mass; cm, circular muscles; dv, dorsal valve; gc, general coelom; gm, gular membrane; hc, head coelom; hl, hinge line; in, interannular nerve; lm, longitudinal muscles; mh, muscles of hinge; mpc, muscles of parapodial canal; oe, oesophagus; pc, paraoesophageal cavity; pca, parapodial canal; ph, pharynx;rs, retractor sheath; sp, septal pouch; vf, ventral foramen; vv, ventral valve. (from Seymour[100])

exceed 16×10^{-2} N cm^{-2}, the sand can support steady pressures of up to 1.5 N cm^{-2}. They suggested that thixotropic properties of the sand-water mixture account, at least in part, for the ability of the lugworm to burrow, but it now seems likely that the frequent scraping action of the buccal mass and pharynx, using little force in each movement, is of prime importance.

The mechanism of extension of the proboscis of *Arenicola* is based upon the cyclical contraction and relaxation of the circular and longitudinal muscles of the body wall and buccal mass,[130] while retraction would appear to involve the sheath of retractor muscles. Eversion of the proboscis does not involve major hydrostatic pressures; it is not thrust out with explosive force but is everted simultaneously with waves of relatively low pressure in the trunk coelom (Fig. 3.5a). The essential feature of probing is a forward-moving wave of simultaneous dilation and elongation of the body wall, followed by a wave of simultaneous thinning and shortening; this ensures progressive anchorage and elongation of the anterior region, dilation of the mouth and eversion of the proboscis against the sand. An additional pressure pulse has further been recorded from the head coelom at retraction. Using an isolate preparation of the anterior segments in which the coelomic pressure could be altered by a manometer-like device, Wells[130] demonstrated that eversion occurs normally at pressures of $2–5 \times 10^{-2}$ N cm^{-2} and that at greater pressures it is irregular and jerky. High pressures, such as those recorded from the coelom later in burrowing (Fig. 3.9), are impracticable during the first stage of the digging period, for the downward thrust can be no greater than the strength of the anchorage. This strength is limited by the light weight of the worm in water and the poor adhesion of an animal of circular cross-section on a flat surface. Anchorage is aided, however, by chaetal extrusion and the erection of flanges which occur from behind forwards, particularly on the anterior three trunk segments (Fig. 3.5a). The anatomy of the flanges is shown in Figure 3.6. Perhaps their most important feature is the parapodial canal, which is an isolated ring of the trunk coelom filled with coelomic fluid. In some very delicate recordings from the parapodial canal, a former colleague, Dr M. K. Seymour,[100] has demonstrated that the flanges are erected by means of a hydraulic mechanism. The pressure in the canal rises with that of the trunk coelom, but is sustained when the pressure of the trunk falls, thereby causing the protraction of the flanges.

Peachia

The infaunal actinarian *Peachia* is phylogenetically remote from *Arenicola*, yet the mechanism of its initial entry into the sand closely resembles the action of the proboscis of the lugworm.[121] The principal modification in *Peachia*, from the normal form of sea anemones, is that the base takes the form, not of a disc, but of a rounded inflatable vesicular lobe, termed the physa. This is primarily a digging organ. When the anemone is laid on the surface of sand, digging begins with the physa turning downward and a series of peristaltic contractions passing down the column. The latter represent direct locomotory waves comparable with the peristaltic waves passing forwards along the trunk of *Arenicola*. The immediate effect of these waves in *Peachia* is not locomotion, but the forcing of the body fluid downwards and the eversion of the physa (Fig. 3.7). Introversion occurs within some 10 s by contraction of the retractor or longitudinal muscles of the column, an eversion-introversion cycle occurring in respect of each peristaltic wave. These waves begin in the upper region of the column at regular intervals of about 1.5 min.[9] Eversion displaces the sand centrifugally, making a cavity

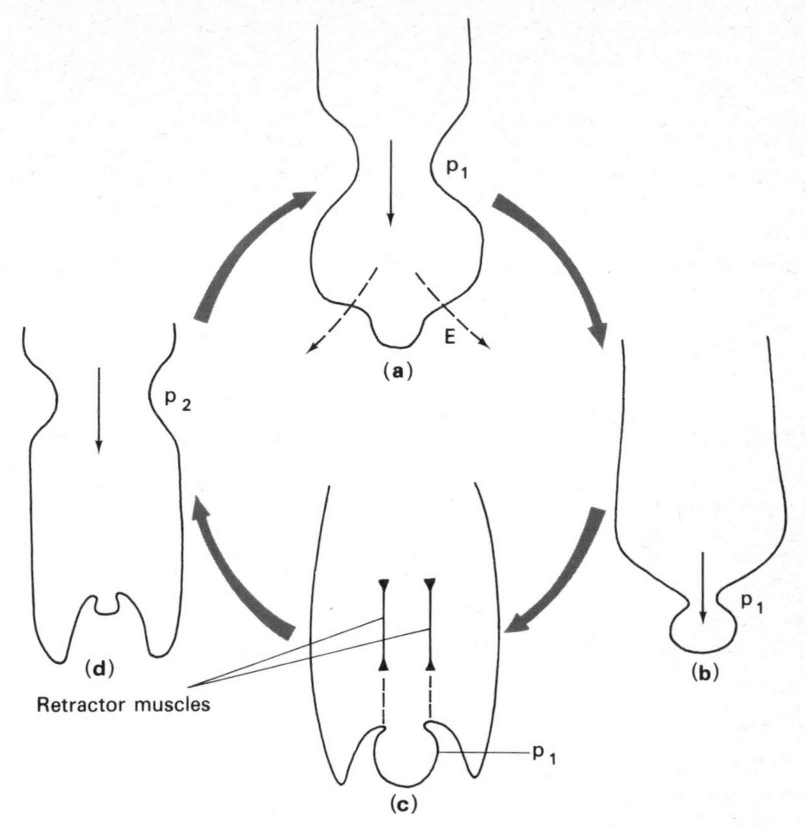

Fig. 3.7 Diagrams of the base of the anemone *Peachia hastata* during the eversion-introversion cycle. (**a**) peristaltic wave (p_1) passing down the column (arrow) causing eversion of the physa and displacement of sand (E). (**b**) further movement of the peristaltic wave to the base. (**c**) introversion of the physa with the contraction of the retractor muscles. (**d**) physa infolded prior to eversion by succeeding peristaltic wave (p_2). (from Trueman and Ansell[121])

into which the column passes. During this initial part of the digging period the area of contact between the column and the sand is limited, as in the lugworm, because of its circular cross-section, and the physa can only press down weakly. But a collar of sand grains has also been observed to adhere to the column in *Peachia* and may facilitate anchorage for initial penetration. However, the first part of the digging period takes about 15 mins for this anemone compared with about 20 s for *Arenicola*. Although the scraping movements are similar, it is clearly less efficient than the lugworm and well illustrates the principle that the circular shape is admirably suited for movement within a substrate but ill adapted for surface locomotion.

Priapulus and Sipunculus

Priapulus, a non-septate coelomate worm, progresses through muddy sand by the successive eversion and introversion of the proboscis,[48] so forming penetration and terminal anchors respectively during this locomotory cycle (Fig. 3.8). The proboscis is forced

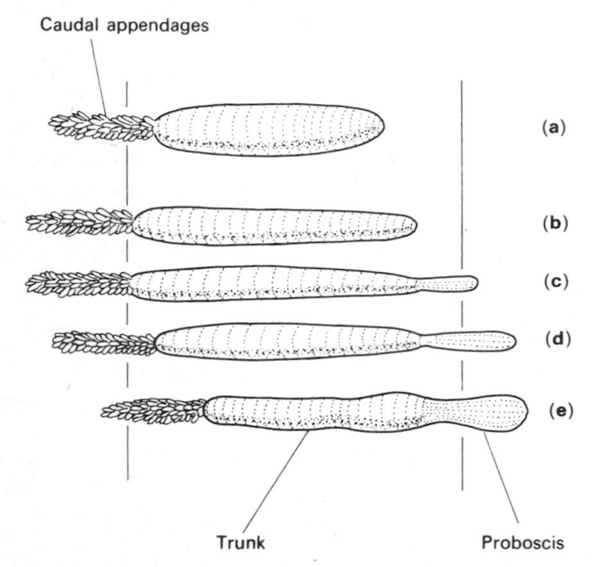

Fig. 3.8 Diagrams illustrating the digging cycle of *Priapulus*. (**a**) Proboscis fully invaginated, body wall dilated to form penetration anchor; (**b**) trunk elongating; (**c**) proboscis thrusting forwards during evagination; (**d**) proboscis commences to dilate to form terminal anchor; (**e**) trunk pulled forward. (after Hammond[48])

directly forwards into the substrate at high fluid pressures by contraction of trunk muscles. When buried, the worm is held in position by dilation of the trunk to form a penetration anchor, but the mechanism is ineffective in respect of entry into the substrate. Often attempts to make an initial penetration of sand are unsuccessful unless the worm encounters some cavity. This is due to the absence of penetration anchorage when on the surface of the sand, and the lack of any specialized device to facilitate initial entry into the sand. A similar state of affairs occurs in *Sipunculus*, where the proboscis is again adapted to movement through the substrate when buried. Whereas the proboscis of *Arenicola* is primarily adapted for scraping a passage through sand, those of *Priapulus* and *Sipunculus* function quite differently. Very much higher coelomic pressures have been recorded in the latter genera in respect of proboscis eversion than in *Arenicola*. These high pressures are clearly incompatible with the beginning of burrowing, although they are highly effective in movement through the substrate once a burrow is established.

All animals that live in soft, often unstable, substrates require at least two abilities to achieve any degree of success. Firstly, they must be able to emerge from the sand when they have been buried too deeply by, for example, wave action; secondly, they should be able to re-enter the sand swiftly when they have been washed out. Many bivalves and polychaete worms are particularly well adapted for re-entry but *Priapulus* and *Sipunculus* are unspecialized in this respect. This may mean that subsequent to initial penetration of the substrate as a juvenile they very rarely emerge, possibly being restricted to well consolidated and relatively undisturbed habitats for this reason.

MOVEMENT THROUGH A SOFT SUBSTRATE

Introduction

The initial part of the digging period may be considered to have ended once penetration has progressed far enough for the medium to sustain the lateral forces applied by a terminal anchor. Once this can be applied the animal is pulled powerfully into the burrow whereas hitherto it has progressed only by relatively weak pushing or probing movements. The manner in which soft-bodied animals burrow may best be considered in detail in respect of four examples, although the mechanical basis upon which various other animals burrow, for example, *Nephtys, Dentalium, Sipunculus, Natica,* is also well established.[121]

Peachia

The principal part of the digging period begins when sufficient of the column, about one-third, is buried for a terminal anchor to be formed by dilation of the physa. More massive contraction of the longitudinal muscles then takes place than occurred during the initial eversion–introversion cycles, and this causes the anemone to be pulled down into the sand. These major retractions, occurring once in each digging cycle, are repeated until burial is complete. The eversion cycle continues between major retractions as direct peristaltic waves pass down the column and, indeed, a small displacement into the sand marks each introversion of the physa. The relationship between the eversion cycle and the movement of head or capitulum of the anemone and the physa are shown diagrammatically for a single digging cycle in Figure 3.18c. Each digging cycle commonly consists of three eversion cycles, the final one causing a major downward movement of the capitulum. The function of the other eversion cycles is to allow the physa to penetrate the sand more deeply. The resistance of the substrate increases with depth, and more eversion cycles then occur during each digging cycle.

During burrowing two principal conditions for anchorage of the anemone occur successively. The condition of *Peachia* closely resembles the generalized diagram (Fig. 3.1) in this respect. Application of the terminal and penetration anchors occurs repeatedly for each eversion cycle, but the dilation of the physa reaches a maximum only once in each cycle when the capitulum is pulled down. A maximum pressure of 5×10^{-2}N cm^{-2} is generated in the coelenteron at this time and in the absence of an oral sphincter muscle the tentacles are retracted. This pressure serves two functions: (i) terminal anchorage and (ii) consolidation of the walls of the burrow. Subsequent contractions of the circular muscles elongate the anemone and, provided the column is anchored, push the physa downwards. This penetration anchor is probably attained partially by dilation of the column above the peristaltic wave and partially by the column possessing an adhesive epithelium by which it may be attached to sand. Very low pressures of 0.3×10^{-2} N cm^{-2} have been recorded in association with eversion of the physa so that only very small forces are being used by *Peachia* in burrowing.[9] Nevertheless, as in more advanced animals, maximal pressure is developed as the terminal anchor is formed and the longitudinal muscles contract to pull the polyp into the burrow. *Peachia* is able to live successfully in a soft unstable substrate by relatively minor adaptations of a very simple body form. The coelenteric fluid is used as the basis of the hydraulic system in the same manner as the coelomic fluid in the lugworm or the blood in the haemocoel of bivalve molluscs.

Arenicola

During the first part of the digging period only minor fluctuations of pressure in the trunk coelom are recorded. Then, after two segments have passed into the sand, high pressure pulses begin and rapidly increase to a maximum of about 1.5 N cm^{-2} as penetration proceeds (Fig. 3.9a). These high pressure pulses characterize the second phase of burrowing by the lugworm and all pressure fluctuations are identical and simultaneous throughout the trunk coelom. One major pulse occurs during each digging cycle (Fig. 3.18b). During the first 1½ min of the second part of the digging period the cycles are all of short duration, but as burrowing continues so the duration of the cycle increases (Fig. 3.9b). Minor

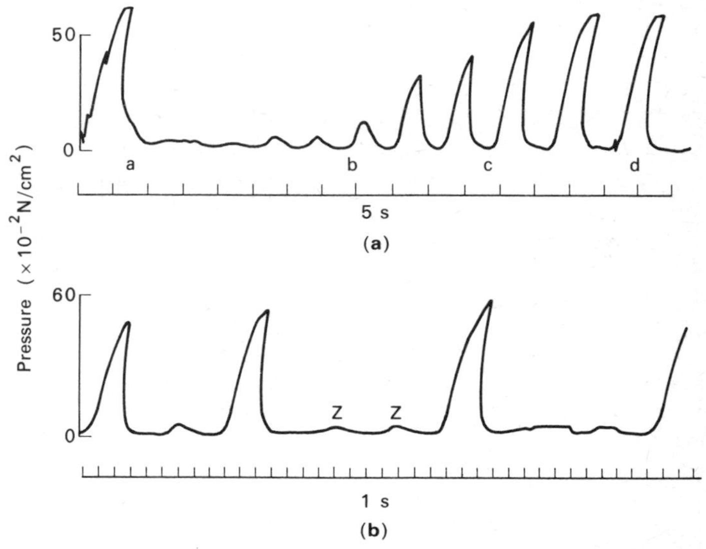

Fig. 3.9 Recordings of the pressures produced in the coelom of *Arenicola marina* during burrowing. (**a**) Pressure changes at the commencement of burrowing; a, head down on sand, followed by repeated proboscis eversion, b, c, and d, 2, 3, and 4, chaetigerous annuli respectively beneath the surface (observed visually when burrowing against glass. The flat top of the last two pulses is due to limitation of travel of the pen. (**b**) coelomic pressure recording after 2 min of burrowing showing increased duration of the digging cycle, minor fluctuations in pressure occurring (Z) between high pressure pulses, corresponding to a peristaltic wave arriving in the anterior segments and proboscis eversion. (from Trueman[110])

pulses corresponding with proboscis eversion then occur between high pressure pulses, one intervening initially, then two, up to a maximum of five. Observations of the lugworm burrowing near the side of a glass aquarium indicated that the high pressure pulse corresponded to dilation of the anterior end of the worm; this forms the terminal anchor (Fig. 3.11) which allows the posterior of the worm to be drawn into the burrow. The effect of pressure in the anterior segments is to cause the body wall to dilate and to adhere firmly to the substrate. This terminal anchor is equivalent to what has been described as an 'anti-sea-gull' reflex. Wells[131] demonstrated this by allowing a worm to burrow down the stem of a filter funnel with the stem closed by rubber tubing clamped together. When *Arenicola* was half-way into the stem pulling the tail caused dilation of the anterior segments, high coelomic pressure and a tenacious grip on the stem.

Each digging cycle of *Arenicola* consists of a major retraction into the substrate, followed by an interval during which the worm elongates and everts its proboscis. At this time the flanges and chaetae function as a penetration anchor to hold the worm in the burrow.[121] It was originally thought that the increase in duration of the digging cycle allowed additional eversion of the proboscis to take place as burial became deeper and the substrate more resistant to penetration. However, more detailed analysis by Seymour[100] has shown that the increasing time interval allows the development of a new activity termed the 'flange-proboscis' (f-p) sequence, in which these two structures are combined as a digging tool. This sequence (Fig. 3.10) is characteristic of the later part of the digging period and a typical digging cycle during this activity is illustrated in Figure 3.18b. The f-p sequence commences with the shortening of the worm (Fig. 3.10f, diagram 1) and in consequence the trunk moves into the sand and the anterior end is pulled back with the flanges raised (2). Eversion of the proboscis (3) scrapes sand away from the end of the burrow, moving the sand outwards and backwards. As the trunk becomes fully shortened (5) the proboscis is suddenly withdrawn, drawing water through the sand into the cavity formed. A recovery phase (6–10) follows with elongation of the trunk, the head penetrating the water space and the sand being liquefied by the inflow of water. The f-p sequence thus has four important functions in burrowing: penetration of the sand, removal of sand from the burrow, softening of sand for further penetration and the drawing of the trunk into the burrow. These movements may be repeated up to five times in each digging cycle. Permanent downward movement may not often occur whenever the terminal or dilation anchor is applied, as it does during the earlier part of the digging period. Movement into the burrow is associated with each f-p sequence and thus may occur between high pressure pulses. This raises the question of the function of the high pressures recorded during the latter part of the digging period if they are not concerned with

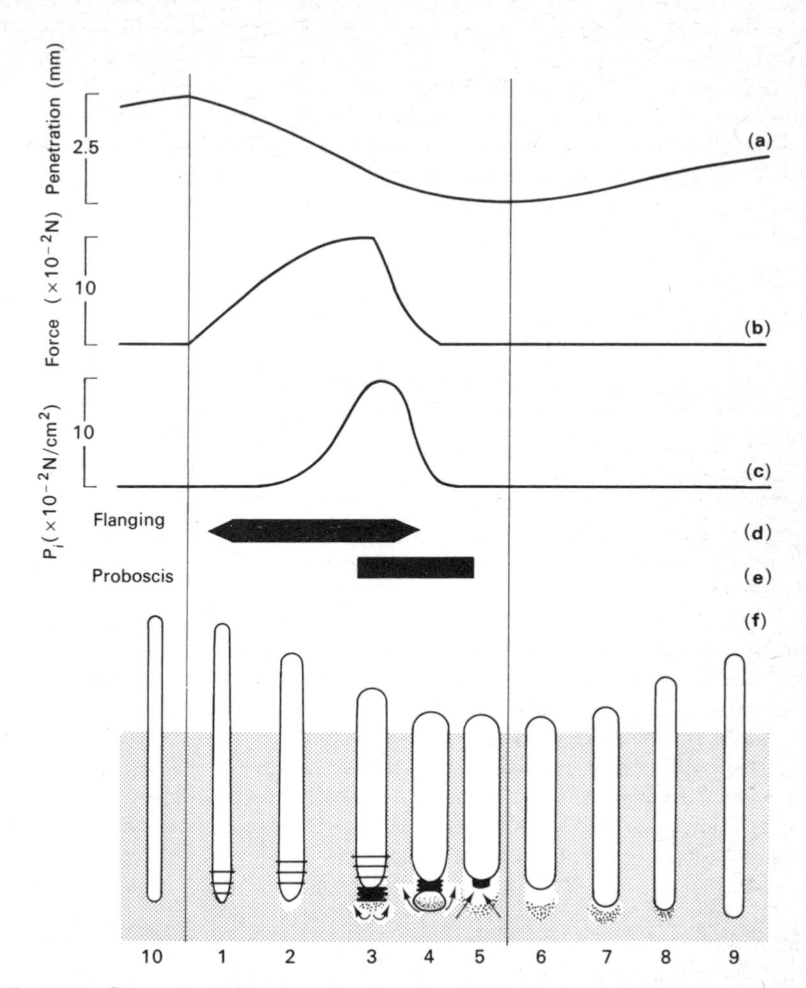

Fig. 3.10 Generalized diagram showing the principal events comprising a single flange-proboscis sequence in the burrowing of *Arenicola*. (**a**) Movement of the posterior trunk into the sand from an isotonic myograph record; (**b**) downward force exerted by the worm; (**c**) internal pressure; (**d**), (**e**) timing of flange erection/depression and proboscis eversion/retraction respectively (from direct observation); (**f**) successive diagrams of a worm at different stages of the flange-proboscis sequence. Large stippled area represents sand; coarse stippling in diagrams 5 to 8 represents the probable volume of sand softened when water is drawn through it in direction of arrows (5) by retraction of the proboscis. The degree of thickening and shortening of the worm is much exaggerated for clarity. (from Seymour[100])

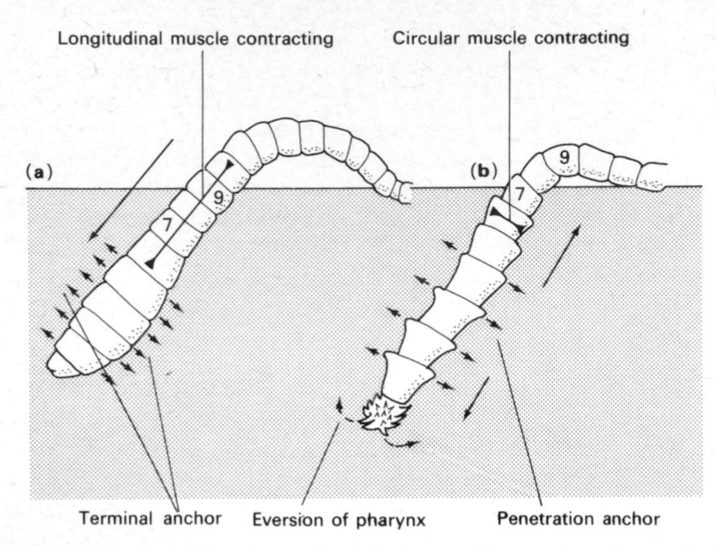

Fig. 3.11 Diagram of the two principal stages of burrowing of *Arenicola*. (**a**) Anterior segments dilated to form terminal anchor; (**b**) penetration anchor formed by flanging. Segments 7 and 9 and movement of the worm are indicated. (from Trueman[111])

anchorage and movement of the posterior trunk into the sand. The most convincing explanation is that lugworms actively enlarge the burrow by high pressure pulses and, with the aid of mucus, consolidate its walls. Indeed, recordings of the pressures produced by *Arenicola* when established in their normal U-shaped burrows, show pressure pulses comparable to those observed during burrowing. This suggests that the burrow wall is being continually maintained.

Lumbricus

Burrowing of the earthworm is outlined here as the basis of comparison between a septate worm and an aseptate one, such as *Arenicola*. Important differences in *Lumbricus* are:

(i) the body consists of functionally separate segments;

(ii) movement is achieved by retrograde locomotory waves;

(iii) *Lumbricus* is terrestrial, essentially burrowing by use of crevices in the soil rather than in a homogeneous sandy substrate;

(iv) the earthworm has no specialized organ of penetration comparable to the proboscis. The internal pressure produced by the shortening of circular muscles provides the force for elongation of the segment against all the forces opposing elongation, which include both internal forces, e.g. stretching of longitudinal muscles, and external environmental resistance. Coelomic pressure also serves to transfer the force of contraction of the longitudinal muscles to stretch the circular ones, and to dilate the segment against the burrow wall. From this it may well be inferred that the coelomic pressure will vary regularly with each locomotory wave.

Pressure recordings from single segments of *Lumbricus* have demonstrated clearly the relationship between pressure and locomotion.[98] At rest, the resting pressures are both remarkably low, between -2 and 2×10^{-2} N cm^{-2}, and free from fluctuations compared with those of *Arenicola*, in which the resting pressure rarely ceases to fluctuate around 5×10^{-2} N cm^{-2}. These differences are attributable to muscle stretch-receptor feedback in virtually the whole trunk body wall. The fully segmented body of the earthworm is clearly more stable and the septate condition allows little distortion of the body under its own weight. The body wall of *Lumbricus* apparently possesses an additional inherent stiffness, which means that pressure is only required for activity involving pronounced changes of shape. The resting body shape can be maintained with complete muscular relaxation, as is indicated by the zero or slightly negative pressures recorded at rest. Similar negative pressures have been recorded during crawling. The outward rebound of the body wall on relaxation of circular muscles, due to subepidermal connective tissue or cuticle, is almost certainly responsible for these values. Indeed the body wall appears to act as a spring; upon circular muscle contraction it stores potential energy, to be released to aid relaxation when the longitudinals contract.

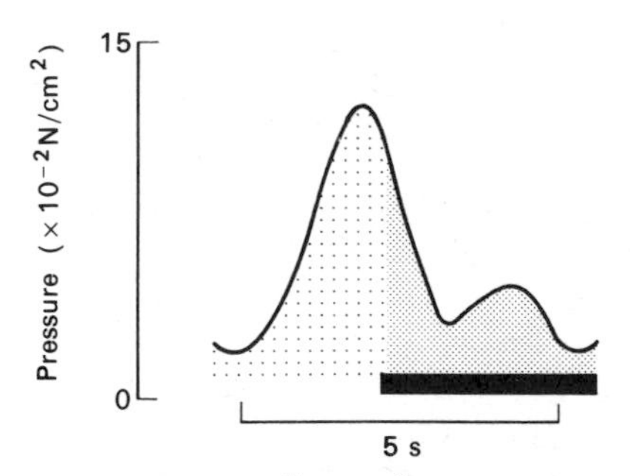

Fig. 3.12 Diagram of pressure ($\times 10^{-2}$ N cm^{-2}) recorded from coelom of single segment of earthworm during locomotion (after Seymour[98]). Coarse stipple represents thinning and elongation of the segment, fine stipple thickening and shortening. Horizontal bar indicates duration of chaetal protraction.

The changes in pressure observed in a single anterior segment are summarized in Fig. 3.12. Maximum force is exerted during elongation of the segment by contraction of circular muscles. This represents the force with which the head can be thrust forward into the substrate. A secondary peak occurs as the longitudinal muscles contract to restore the circular muscles and to achieve maximum segmental thickness so as to produce a *point d'appui*. Chaetal protrusion is simultaneous with this. Simultaneous recordings of pressure 26 segments apart demonstrates the passage of a retrograde wave (Fig. 3.13). Maximum pressure occurs in both segments

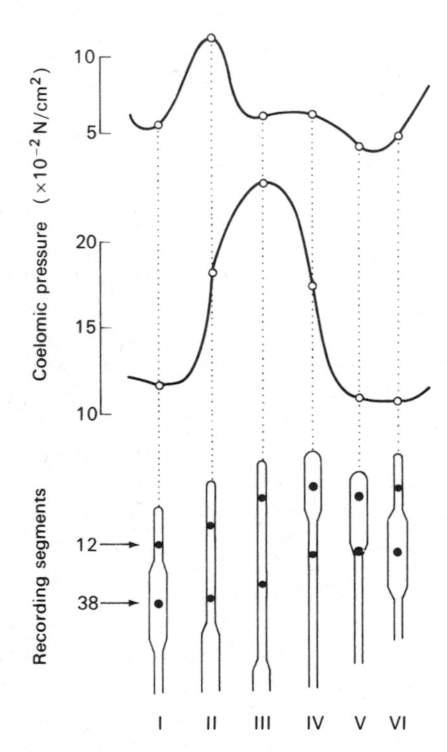

Fig. 3.13 Simultaneous records of coelomic pressure of *Lumbricus* in segments 12 (above) and 38 (centre) during the passage of a locomotory wave resulting in forward locomotion. The cannulae were free to move with the worm, which was crawling over damp earth. Diagrams of the anterior half of a worm showing the shapes and relative position of the recording segments at various stages of the locomotory cycle are shown below. Stages I to III show the thinning phase passing back from the anterior end, and IV to VI the thickening phase following it; stage I is the same as stage VI except that by the latter stage the worm has moved forward by one step of 2–3 cm. (after Seymour[98])

during elongation, although the amplitude of the pulse differs. This indicates that the musculature of the worm may show different levels of activity in the same locomotory sequence. Shortening of segment 38 by contraction of longitudinal muscles does not produce a pulse, a condition apparently characteristic of the more posterior segments when crawling over the surface of the soil. However, when the animal is burrowing the longitudinals play a much more important role. They must pull the more posterior segments into the soil and dilate them to enlarge the burrow which is being formed. These functions are very similar to that of the terminal anchor in *Arenicola*. In both worms high pressures are primarily required for this dilation. In *Lumbricus* this may be limited at any one instant to a relatively small number of segments. Seymour[98,99] has calculated that the maximum 'burrow packing force' that *Lumbricus* can generate by longitudinal muscle contraction is about 3.75 N cm^{-2}.

Bivalve molluscs

This group of animals is particularly well adapted for an infaunal mode of life and exhibits many specializations. The burrowing habit has been more extensively investigated than in other classes because of the diversity of species, and the facility with which transducers can be connected to the shell. It was additionally the first group whose burrowing activity was described using modern electronic techniques.[122]

The description that follows is of the burrowing of a generalized dimyarian bivalve, for this activity is remarkably similar throughout the class. There are only some relatively minor modifications in anatomy in relation to shell shape and habit.

Some aspects of the anatomy of a dimyarian bivalve must be outlined before discussing the results of experimental work. Apart from the foot which is protracted during locomotion, the body is enclosed in a shell consisting of two lateral valves and an elastic ligament (Fig. 3.15). The valves are drawn together by paired adductor muscles and part of their force is stored in the ligament as potential energy to cause the valves to gape. When extended for locomotion the foot is relatively large and consists of two parts, dorsally a viscero-pedal region, and ventrally a muscular region into which the haemocoel extends. The latter part is characteristically compressed and blade-like, being adapted for rapid penetration of the sand as in the Tellinacea. The pedal musculature essentially consists of three pairs of shell muscles, the anterior and posterior retractors and the protractors (Fig. 3.14), and the transverse muscles (Fig. 3.15). Retractor muscle fibres spread through the ventral part

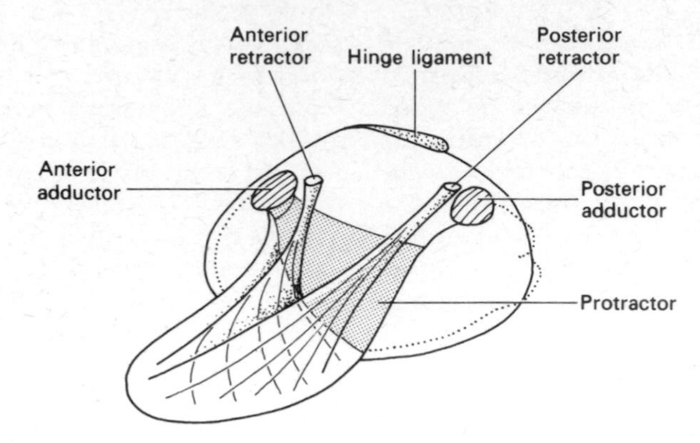

Fig. 3.14 Sagittal section of a generalized bivalve with foot extended showing principal musculature used in burrowing.

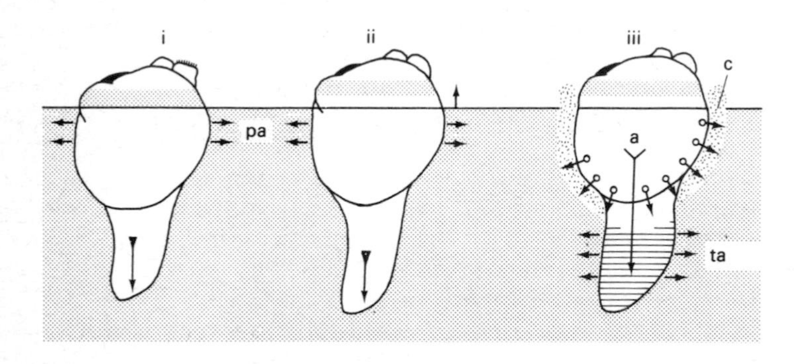

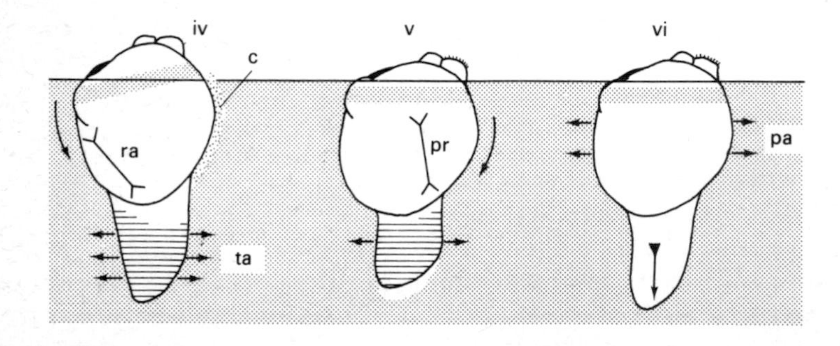

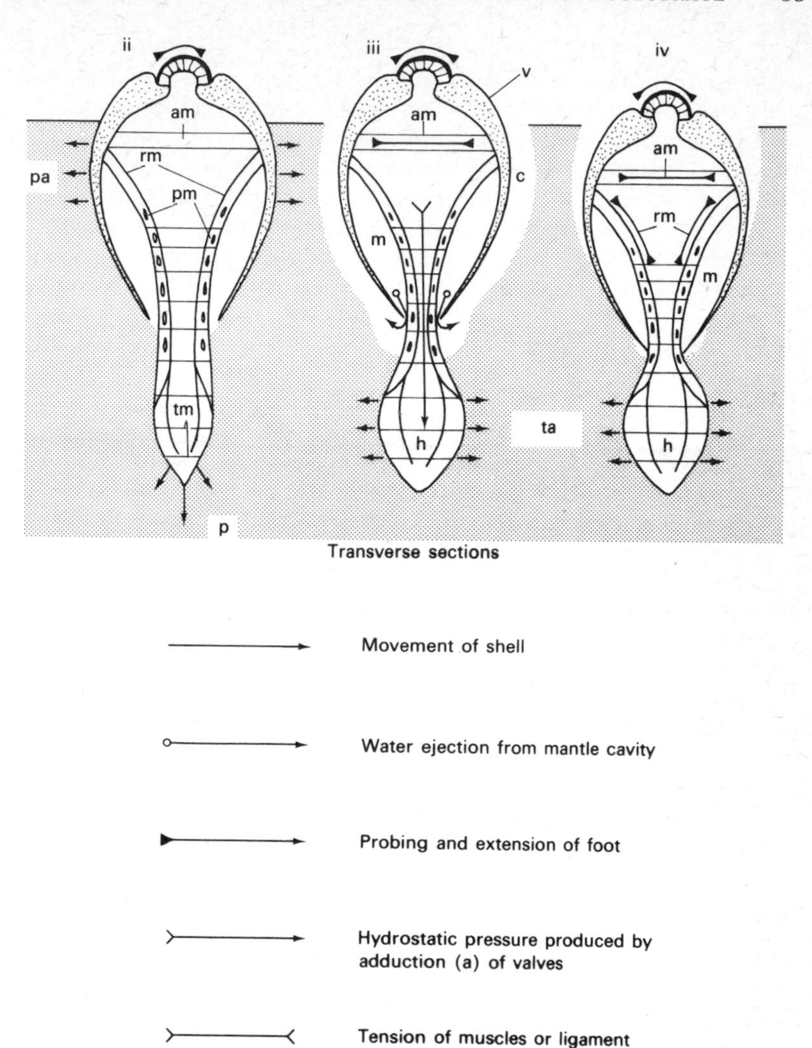

Fig. 3.15 Diagrams of successive stages (I–VI) in the burrowing of a generalized bivalve showing penetration (pa) and terminal (ta) anchors (arrowheads). The transverse sections correspond to the stages above. Dotted band across the valves indicates shell movements with reference to the sand (stipple) surface and the horizontal shading on the foot the region of the pedal anchor. am, Adductor muscle; c, cavity formed in sand by water from mantle cavity; h, pedal haemocoele; m, mantle cavity; p, probing by foot; pm, protractor muscles; pr, posterior retractor; ra, anterior retractor; rm, retractor muscles; tm, transverse pedal muscles; v, valves. (after Trueman[116])

of the foot forming a complex trellis-like or geodesic network. Muscle fibres so arranged are not mutually antagonistic. They achieve postural control of the foot by the shortening of one or other set of fibres but additional antagonistic muscles, for example transverse muscles, are required for lengthening. Apart from the protractor muscles, which act as circular muscles around the upper part of the foot, there are little or no circular muscle fibres such as are present in the Annelida. The retractors are opposed, in place of these, by intrinsic transverse muscles passing across the haemocoel between the connective tissue of the basement membrane of the pedal epithelium. The blood in the pedal sinus, restricted from flowing out of the foot by Keber's valve, functions as the fluid of this fluid-muscle system. Only in the razor shells (*Ensis*) does the anatomy of the foot differ markedly from this description. Here the foot is circular in cross-section, retraction being accomplished only by the posterior retractors, for the anterior retractors are modified to function in a similar manner to the protractors.

After penetration of the sand has been achieved by the foot and the shell drawn erect, the second part of the digging period begins. This consists of a series of digging cycles (Fig. 3.4), each of which involves a number of co-ordinated activities repeated in the same sequence for each cycle. Both the pattern of the digging period and the sequence of the digging cycle are common to almost all burrowing bivalves, e.g. *Nucula, Glycymeris, Anodonta, Cardium, Tellina, Donax, Venus, Mactra, Mya* and *Ensis*.[116]

The digging cycle is best understood by reference to Figures 3.15, 3.17 and 3.18a, which are derived from analysis of cine film and recordings. It consists of the following stages, numbered i–vi to correspond to other descriptions.[116]

(i) The foot makes a major probe downwards, tending to raise the shell if penetration is not easily achieved.

(ii) The siphons close so as to prevent water from passing out at adduction; the probe continues to maximum pedal extension and dilation of the foot may begin.

(iii) Rapid adduction of the valves occurs, causing the ejection of water from the mantle cavity through the pedal gape. It also generates in the haemocoel a pressure pulse which is responsible for dilation of the foot to form a terminal anchor.

(iv) Contraction of first the anterior and second the posterior retractors, resulting in the shell being drawn into the sand. The siphons reopen at or just prior to the end of retraction.

(v) Adductors relax, and the shell opens by the energy stored in the ligament so as to press against the sand and form the penetration anchor.

(vi) A period in which the shell is static and the foot is protracted by a repeated probing action. This is equivalent to the plateau in traces of entire digging periods (Fig. 3.4).

Digging cycles consist essentially of two phases which occur successively. First, penetration anchorage, by the shell being pressed open against the sand so that the foot may extend. Second, adduction followed by retraction, which has the converse effect, for the foot is firmly pressed against the sand while the shell is drawn down. This is effectively the alternation of penetration and terminal anchors in a manner similar to that described for a generalized burrowing animal (Fig. 3.1).

Anchorage of the foot is achieved by the pressure pulse derived from the force generated by the sudden adduction of the valves (Fig. 3.16a). This may produce a broad flat area of contact with the sand as in *Donax* or *Anodonta*, a bulbous swelling as in the tubular foot of *Ensis*, or the outward spreading of a cleft foot as in *Glycymeris*. The effectiveness of the foot as a terminal anchor depends on its ability to apply sufficient force to compact the sand laterally and is related to three factors: the weight of sand directly above any swollen outwardly-projecting area, the effect of suction beneath the foot, and the frictional effect of adhesion of sand to the foot by mucous secretion.

Adduction of the shell valves destroys the penetration anchor and has two further functions. It forces a jet of water from the mantle cavity into the sand around the proximal part of the foot and valve margins. At the same time blood is forced into the distal region of the foot at high pressure (Fig. 3.16) so as to provide the force for the formation of the terminal anchor (Fig. 3.17). Downward motion of the shell or pedal retraction is facilitated both by the reduction of the profile of the shell and the production of a fluid sand-water mixture around the shell. The direction of the water jet into the sand is controlled by the opposing edges of the mantle edge. The volume of water expelled has not been measured directly but observations of the angular change between the valves at adduction and their surface area have been used to make an estimate of the reduction on the volume enclosed by the valves. For *Ensis arcuatus* the angle of closure is 20°, the reduction in volume being about 6.25 ml, or 20% of the volume enclosed between the valves. 10–15%, however, as estimated for *Mactra corallina*, is probably a more typical value. This includes both water from the mantle cavity and blood flow into the extended foot, although the jet of water undoubtedly represents the greater part. For example, in *Ensis* the foot increases in volume by 2 ml, so that more than 4 ml of water is expelled.[113]

Downward movement of the shell is brought about by the contraction of the proximal regions of the retractors. Anterior

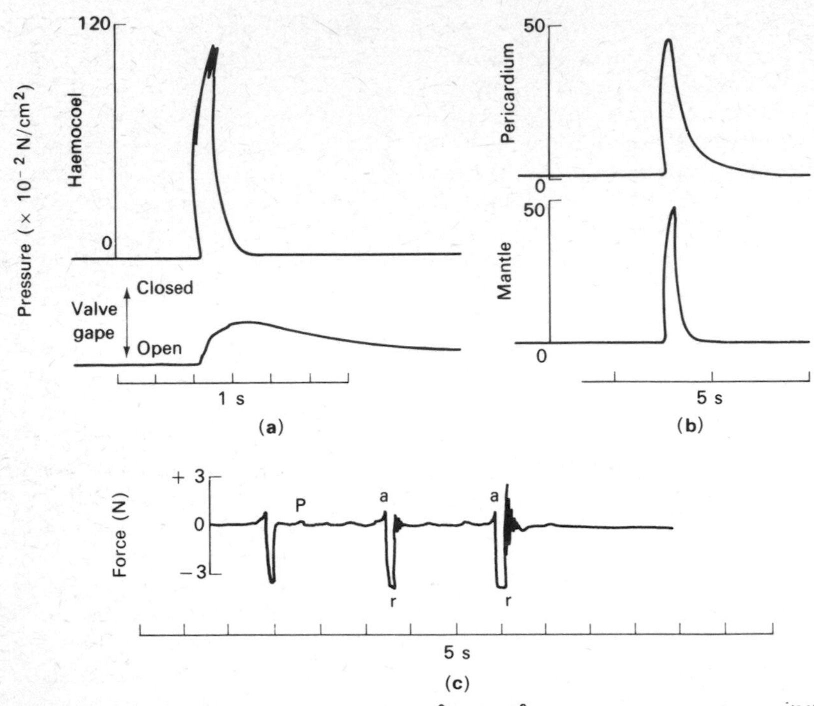

Fig. 3.16 Recordings of pressure ($\times 10^{-2}$ N cm^{-2}) (**a** and **b**) and of forces (N) generated by bivalves during burrowing (**c**). (**a**) pressure in pedal haemocoel (above) produced by adduction of valves (below) of *Ensis arcuatus*; (**b**) pressure pulses of similar amplitude but different duration in pericardium and mantle cavity of *Margaritifera margaritifera*; (**c**) force (N) of probing (P) and pedal retraction (r) of the foot of *E. arcuatus* measured as described in text. a represents maximum amplitude probe occurring briefly at adduction. (**a** and **c** from Trueman,[133] **b** from Trueman[112])

retractor muscles always contract before the posterior, imparting a single rocking motion to the shell in each digging cycle, but in some species of *Mactra* this motion is repeated. Rocking motion is most characteristic of bivalves with fat shells, the see-saw action enabling them to penetrate the sand obliquely rather than directly. Many rapid burrowers, for example *Donax*, penetrate sand in the direction of the long axis of their shell with the foot protruding anteriorly. In these animals rocking is suppressed and posterior retraction attains greater importance. Fullest development of this modification of the basic bivalve pattern is shown in such specialized burrowers as *Ensis* or *Solemya*, whose shells are greatly extended anterio-posteriorly.

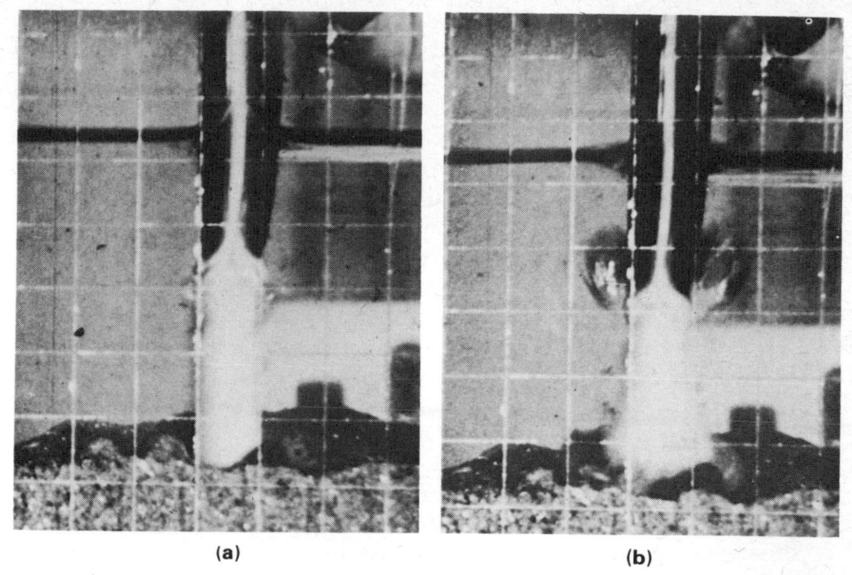

Fig. 3.17 Photographs of the shell and foot of *Ensis* (razor shell) taken at an interval of 1/16th s during a digging cycle carried out while held in water. A 1 cm grid is superimposed. (**a**) Pedal probing with the shell gaping as indicated by mantle between valve margins; (**b**) adduction of valves, dilation of foot and expulsion of water jets from mantle cavity.

Movement of the shell into the substrate is determined by the strength of the retractors, by the shell shape, and by the resistance of the substrate. Retractor strength has been determined for a number of bivalves by attaching a force transducer to the shell by means of a thread. The maximum force recorded during retraction is generally equivalent to a muscular tension of about 15 N cm^{-2}, retraction strength being directly dependent on the cross-sectional area of the retractor muscles. Those bivalves, for example *Ensis arcuatus*, with powerful muscles have the most effective pedal (terminal) anchorage and move rapidly into the sand. Smaller muscles and poor terminal anchorage are characteristic of animals with heavy, broad shells, for example *Mercenaria mercenaria*. Indeed these latter animals commonly burrow slowly, largely by dropping into the cavity formed in the sand by the water jet from the mantle cavity, rather than by pedal retraction. The ratio between the strength of retraction and weight in water serves as a simple comparison between bivalves, values of 100 and 0.25 respectively being recorded for *E. arcuatus* and *M. mercenaria*.

Nair and Ansell[80] have attempted to assess quantitatively, the effect of shell shape and substrate resistance on burrowing. This they have done by comparing the movement per unit of applied force for dead shells with the force developed by the retractors during burrowing, recorded by attaching a force transducer to the shell by means of a thread. In bivalves showing reduction in breadth of the shell, facilitating movement through the substrate, the movement of dead shells per unit applied force is similar to that derived from the retractor muscles during burrowing, for example in *Tellina*. In those with more tumid shells, for example *Mactra*, the movement during burrowing may be up to 75% greater in life than with the same force applied to a dead shell. This is accounted for both by the rocking motion and by the water jet.

Whilst the shell is static, forming a penetration anchor, the foot is extended with a series of probing movements. Characteristically the time spent in probing increases during the second part of the digging period (Fig. 3.4a). In any individual bivalve the probing rate is constant, although this may vary both with size intraspecifically and between species, e.g. *Ensis arcuatus*, 90 probes/min; *Mercenaria mercenaria*, 16; *Mya arenaria*, 1. In any single animal with a constant probing rate, any increase in the duration of application of the penetration anchor (stage (vi) of the digging cycle) results in an increase in the number of probes per cycle. This results from the substrate becoming more resistant with depth.

COMPARATIVE SURVEY OF BURROWING

Some comparison of burrowing is attempted here using the examples given above as a basis. In almost all soft-bodied animals the digging period is clearly divisible into two parts, initial penetration and extension into the substrate. Only weak penetration forces may be available during the first, and many animals show specific adaptations to overcome this difficulty, for the ability to re-enter the substrate immediately when they have been dislodged by waves, for example, is an important factor in the survival of infaunal animals. These adaptations include the use of cilia in *Natica* (Fig. 3.19) and the hemichordate *Saccoglossus*, an eversible proboscis as in *Arenicola*, or rapid undulatory swimming towards the sand as in *Nephtys* or as in amphioxus.[121]

Details of the digging cycles of the second part of burrowing have been described for some animals, and these are summarized for easy comparison (Fig. 3.18). Each shows the alternation of terminal and

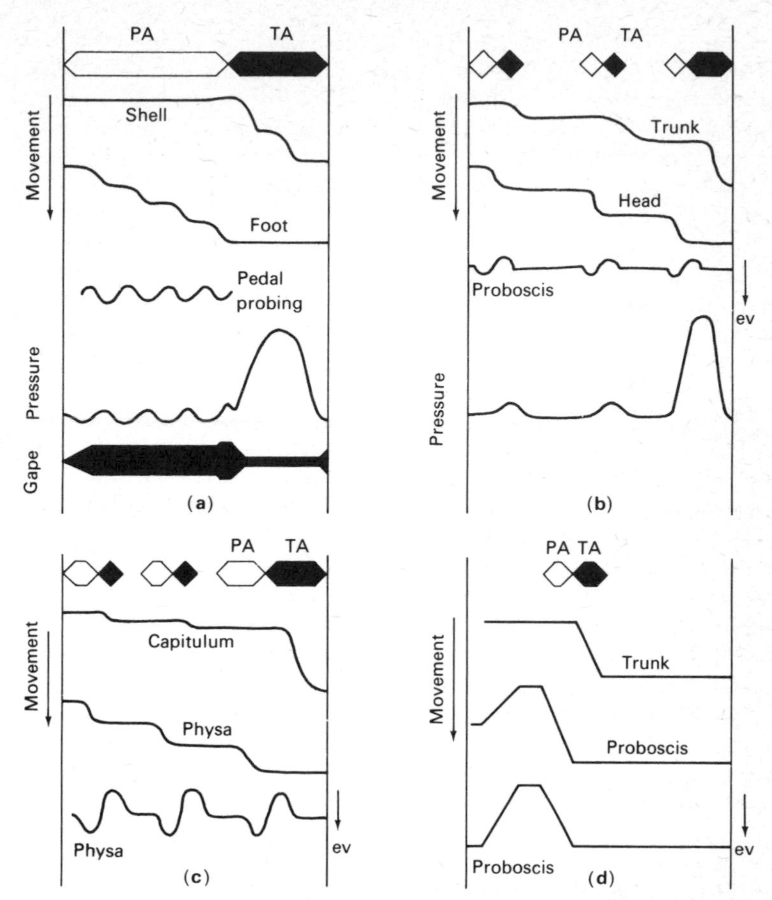

Fig. 3.18 Graphical survey of the events of digging cycles of (**a**) a generalized bivalve mollusc; (**b**) *Arenicola*; (**c**) *Peachia*; (**d**) *Priapulus*. (**a** and **c** after Trueman and Ansell,[121] **b** after Seymour,[100] **d** after Hammond[48]) Each diagram shows the periods when penetration (PA) and terminal (TA) anchors are applied; the movement of the anterior and posterior regions of the body and the introversion–eversion (ev) or probing cycle as labelled. In (**a**) and (**b**) pressures in pedal haemocoel and trunk coelom are indicated respectively. In order to facilitate comparison this diagram is not to scale. Detailed descriptions are given in the text.

penetration anchors which is basic to all soft-bodied forms. Only when one or other is applied can forward motion be achieved; thus one part of the body must be fixed to permit another region to thrust forwards. Effectively, this is a process of stepping, not unlike looping in a leech, where the whole body is involved in each step. However, because of the resistant medium encountered in burrowing, the step length is generally only a small portion of the body length. Essentially the principal of anchorage of part of the body to sustain backthrust while another part extends is common to all forms of locomotion by all animals, but the principal can nowhere be seen so clearly as in the movement of soft-bodied animals through soft substrates.

In most burrowing animals high pressure pulses cause dilation of part of the body to form the terminal anchor and to consolidate the burrow walls. These pulses are developed at least in part by the contraction of the longitudinal muscles, together with local relaxation of circular or transverse pedal muscles to allow dilation of the body. Exceptions are *Sipunculus* and *Priapulus*, where high pressure is generated for proboscis eversion. These are representative of two groups of non-septate coelomates in which the proboscis cavity forms part of the general body cavity and muscular tension applied posteriorly can affect the proboscis. In the absence of specialized scraping devices, the proboscis is shot out with great force by the almost simultaneous contraction of circular and longitudinal muscles of the trunk. Thus any penetration anchor formed by dilation is lost because of reduction of trunk diameter. Anchorage is maintained, however, by flexure of the body so as to grip the burrow. After extension of the proboscis the body is drawn forward by the contraction of the proboscis retractor muscles. A similar method of locomotion in which the extrovert retractor muscles again provide the main tractive force is also encountered in burrowing Nemertea, e.g. *Cerebratulus*, but in this group the fluid-filled cavity is the rhynchocoel.[121] In contrast, in most burrowing polychaetes the proboscis retractor muscles do not provide traction, the proboscis functioning, as in *Arenicola*, only to excavate a hole into which the animal extends. The manner in which the proboscis of *Nephtys* is violently everted to punch a hole in the sand while the body is held firm against the burrow wall by the parapodia and chaetae of segments 15–50 is fully described by Clark and Clark.[27] The prostomial horns of *Polyphysia* fulfil a similar function to the proboscis of *Arenicola* in scraping out the substrate, but the trunk advances by means of a series of direct peristaltic waves without any clear pattern of penetration and terminal anchorages. Only low pressures are developed by this polychaete in using direct waves as a means of locomotion and Elder[34] points out how well this adaptation fits in with life in soft muddy substrates. Elder[35] also observes how tentacles of a burrowing apodous holothurian, *Leptosynapta*, are used in a similar manner to the horns of *Polyphysia* to excavate the sand, while the perivisceral coelom acts as a fluid skeleton for the body wall muscles when moving through the burrow. In this respect these two genera show strikingly convergent evolution in a common type of habitat, although members of very different phyla. Enteropneusts, on the other hand, for example *Saccoglossus*, have a quite different burrowing habit from other coelomates in so far as the proboscis is the most active part of their body, the remainder of this worm-like animal being dragged passively behind.[60]

In all Mollusca the common step-like locomotory pattern of alternating terminal and penetration anchors obtains (Fig. 3.18a), but the Bivalvia are unique in employing a double fluid muscle system whereby the substrate around the shell can be liquefied at the same instant as the terminal anchor is applied. This is one advantage of the bivalved shell. Yet another is the use of the elastic ligament to serve as an energy store effecting penetration anchorage. Indeed the molluscan shell is so well adapted by the Bivalvia to the burrowing mode of life that it would appear likely that they originated as infaunal animals. The Bivalvia is not the only molluscan class to inhabit soft substrates successfully, other notable examples being the scaphopod, *Dentalium*, and the gastropods *Natica*[121] and *Bullia*.[8] In all molluscs that burrow, the foot is used to achieve penetration and to serve as a terminal anchor, but except in the

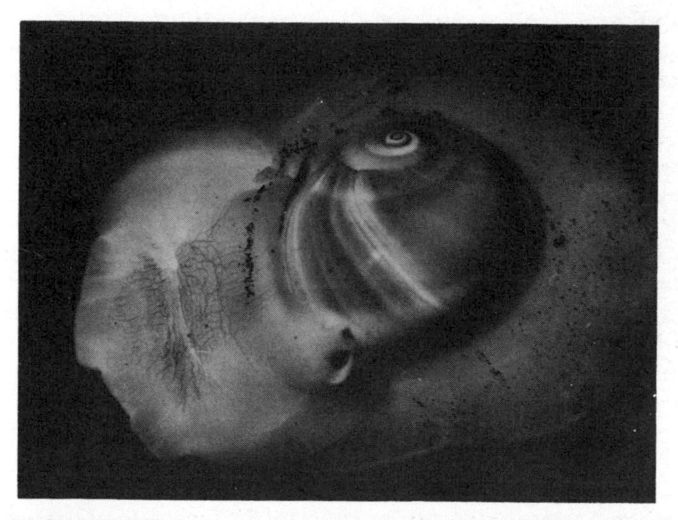

Fig. 3.19 Sand dwelling gastropod, *Polinices (Natica)*, showing enlarged foot with propodium on left forming a cephalic shield used for burrowing through sand. Overall length 8 cm.

bivalves only intrinsic pedal muscles are utilized to effect changes in shape. It is noteworthy that in some members of the Naticidae, e.g. *Polinices* (Fig. 3.19), a large quantity of water is taken in during expansion of the foot to supplement the blood in its hydraulic role.[94,115] One of the earlier workers in this field, Schiemenz,[96] described the foot of *Polinices* and the role of water in its expansion. It was unfortunate that *Polinices* is an exception in this respect for Schiemenz's investigation led to a false belief that water had an important part to play in the expansion of the foot of many molluscs.

In addition to their ability to penetrate and to progress through soft substrates, burrowing animals must be able to emerge from the substrate when covered too deeply (by wave action, for example) if they are to be successful in this habitat. This is achieved in worms by simply turning upwards and progressing until contact with the surface of the sand is attained. By contrast, bivalve molluscs often emerge by pushing downwards with their foot so as to thrust the shell up. This is accomplished by adduction occurring with the foot partially withdrawn, as may be observed if an *Ensis* is held in the hand and caused to close the valves. A similar mechanism is used in some cockles to leap above the substrate as an escape reaction.[5]

Little reference has been made to the burrowing of hard-bodied animals, such as brittle stars and crabs. In general, they have multiple appendages, with some of which they can scrape a hole whilst others afford anchorage or propel the animal forwards. The only detailed analysis has been made in respect of the mole crab *Emerita*,[119] which burrows into the sand backwards. The 4th pair of thoracic limbs and uropods are specialized to excavate the burrow, whilst the first three pairs of legs propel the crab into the sand with a motion similar to rowing. This division of the locomotory apparatus into two functional groups is comparable to what is found in soft-bodied animals. The proboscis in *Arenicola* or the physa in *Peachia* achieve penetration while progression, achieved in them by terminal anchorage and contraction of longitudinal muscles, is equivalent to the action of the first three thoracic limbs. In comparison with soft-bodied animals, multiple appendages allow much greater specialization of function. Although removal of one pair of either group of limbs did not prevent burial, the inability of *Emerita* to burrow with all limbs of either group amputated indicated that their functioning was not sufficiently plastic to allow this degree of adaptation.

A general comparison of the mechanisms used for burrowing in different phyla has been given above in purely descriptive or qualitative terms. A truly comparative assessment of the ability of different animals to burrow can only be made in quantitative terms. Nair and Ansell's[80] comparison of the force required for burial by bivalves of different shell shape is a valuable preliminary for such a study, but for a quantitative comparison between all animals, the energy required for burial must be determined.

Although such determinations are practicable, using either force and displacement transducers, or oxygen determination during burial, our researches are not yet sufficiently advanced in detailed analysis except in respect of two species of the clam, *Donax*,[10] these being migratory species found on tropical beaches. The analysis of their locomotion indicates a possible approach to ascertaining the energy requirement for burrowing. The West Indian species, *D. denticulatus*, moves into the substrate within 4 s in a rapid series of digging cycles. The power required for movement of an animal may be expressed as the drag (*D*) multiplied by the velocity (*DU*).[2] This determination has previously been applied to burrowing in respect of the mole crab *Emerita*.[119] Drag cannot be measured directly, but it is possible to record the maximum force exerted in burrowing whilst the animal is restrained by a thread attached to a force transducer. This represents the maximum locomotory force that the animal can exert and normally it must overcome drag forces in burrowing. Hence drag cannot exceed the force recorded and this may be assumed to represent a maximal value for drag. *D. denticulatus* of 2 cm length developed a maximum force of 0.1 N and a mean velocity of burial of 0.5 cm s^{-1} so that the power required is $5 \times 10^{-4} \text{ J s}^{-1}$ (= watts). In this species complete burial is achieved in 4 s with a total energy requirement of 2×10^{-3} J equivalent to 0.48×10^{-3} cal (1 cal = 4.18 J). If 5 cycles are required for burial, the energy requirement for each digging cycle is about 1×10^{-4} cal (4×10^{-4} J approx.).

FLUID DYNAMICS OF BURROWING

Introduction

Wherever burrowing is successfully accomplished by soft-bodied animals a large fluid-filled cavity is used. In general the larger the cavity the more powerful the digging action. This is accounted for both by the amount of fluid allowing changes in body shape of greater magnitude and by more extensive body cavities facilitating the participation of more body muscles in the burrowing activity. In different phyla cavities of different origin serve as the basis of the hydraulic system, as for example the coelenteron in Coelenterata, the coelom in aseptate (Sipunculida) and septate (Annelida) worms, the rhynchocoel in Nemertea, and the haemocoel in Mollusca. Use of the coelenteron for a hydrostatic function as in anemones has the obvious disadvantage of mixing the hydraulic and the alimentary function, for the mouth must be closed during changes in shape. The haemocoel allows the gut to be freed from the effect of changes in shape of the body wall while the coelom additionally separates hydrostatic and circulatory systems. Both the latter are more successful hydraulic systems than the coelenteron, but all function in a similar manner and exhibit maximal development in respect of the burrowing habit. It may be argued that both haemocoel and coelom served primarily as hydraulic organs which evolved as adaptations to the burrowing habit.[26] Attention can now be appropriately given to their function in bivalve molluscs and worms.

Bivalve molluscs

Extension of the foot, its probing into the sand and the initial stages of dilation to form the terminal anchor are brought about by intrinsic pedal muscles and the blood in the haemocoel. Protraction is caused by contraction of transverse and protractor muscles with the relaxation of the retractors (Fig. 3.15). The converse condition with the retractors under tension, alternates with the former during the probing action. To allow this antagonistic action the pedal blood must be maintained at constant volume; it is therefore prevented from leaving the foot by Keber's valve, situated between the pedal haemocoel and afferent vessels to the gills. In general, only low pressures are developed in the haemocoel during probing, although as much as 10×10^{-2} N cm^{-2} was recorded in *Ensis.*[113] It is always difficult to record pressures from within the foot; an alternative approach, that of measuring the force of probing, has therefore been attempted with *Ensis*. Specimens were held vertically by a clamp so as to prevent vertical movement of the shell while allowing adduction and pedal movements. *Ensis arcuatus* (13 cm long), so arranged, dug actively into a large beaker of sand and water carefully balanced on a beam balance. A force transducer was attached to the pan by a wire and allowed to take up a residual load. This effectively gave a zero level, about which any pushing or pulling on the sand gave respectively positive or negative deflections (Fig. 3.16c). Maximum force at retraction was more than 10 N, equivalent to a tensile force of 20 N cm^{-2} in the posterior retractors. The force at probing varied but values of 0.4 N were commonly obtained. Assuming that a cylindrical foot of 2 cm^2 cross-sectional area was thrusting down on the substrate, the thrust would be associated with a haemocoel pressure of 20×10^{-2} N cm^{-2}. This value, which is rather high when compared with recordings using pressure transducers, could be accounted for either by a greater area of the foot being in contact with the substrate than used in this estimation or by the difficulties inherent in recording these pressures from within the foot during probing. Some subsequent recordings of the pressure within the foot of *Anodonta*[19] or *Donax* of $20–40 \times 10^{-2}$ N cm^{-2} during probing suggest that the pressures deduced from the experiment with *Ensis* on the beam balance are probably correct.

Recordings of pressure from the mantle cavity, pedal haemocoel or pericardium show simultaneous pulses at adduction, the pressure pulse affecting the entire body. The foot, shell and body musculature can be thought of as a hydraulic system in which the force of contraction of the adductors can be transferred to the foot to cause dilation and terminal anchorage. Retraction immediately follows adduction and sustains the pedal pressure. Multiple pressure pulses in which the final peak corresponds to retraction have been recorded in *Ensis* (Fig. 3.16a). Blood is retained in the foot by Keber's valve, which thus ensures that dilation and anchorage is maintained until the end of retraction.

The fluid muscle system of a bivalve is characteristically a double system consisting of mantle cavity and haemocoel. Adduction affects both simultaneously, for pressure peaks of nearly equal amplitude can be recorded during digging from both cavities in a variety of bivalves[19] (Fig. 3.16b). Where the opposing mantle folds are free, water can escape from around the ventral margin of the mantle cavity, as in the fresh water mussel *Margaritifera*, and the duration of the pulse in the mantle cavity is markedly less than in the pericardium. Keber's valve is important in preventing a surge of blood from the foot into the kidney and gills during the interval when pressure is low in the mantle cavity but high in the foot. In *Ensis*, however, the ventral mantle margins are extensively fused so that the mantle cavity is almost completely enclosed and the peaks are of similar duration.[116]

The muscle tensions developed in the production of high pressure pulses have been estimated for *Ensis*.[113] This involved the measurement of the cross sectional areas of all the muscles whose contraction draws the valves together (i.e. the adductor and other cross-fused mantle muscles) and their distance from the hinge axis of the shell. The application of a moment (M) about the hinge axis of a valve of dorso-ventral height (h) produces a pressure over the surface area (A) of a valve. The relation between applied moment and hydrostatic pressure may be expressed

$$M = PA \cdot \tfrac{1}{2}h$$

If the adductor muscles alone are considered, then for a pressure pulse of 1 N cm^{-2} they must develop a tensile force of 50 N cm^{2} if due allowance is made for opposing the opening movement of the ligament. The adductors only represent about 36% of the total muscle fibres extending between the valves of *Ensis* and if all fibres participate in adduction then the tensile force required is reduced to 15 N cm^{-2}. This is a similar value to that obtained for the contraction of the retractors, so that pressures of 1 N cm^{-2} are clearly well within the limits of the muscular strength of the razor shell.

Annelid worms

Arenicola is a segmental coelomate in which the intersegmental septa have been secondarily lost in the trunk so that the long trunk coelom is functionally equivalent to the coelom of unsegmented coelomates such as *Sipunculus* or *Priapulus*. It can be shown in *Arenicola* that pressure pulses in the trunk do not affect that in the septate tail. Septa between segments, as in the earthworm (p. 60), divide the body into mechanically nearly independent units which allow localization of reciprocal muscle action, so giving more effective locomotory organization. However, a single large body cavity has two advantages which must have been important to account for the loss of septa in *Arenicola*. Firstly, the coelomic fluid can act as a hydraulic system so that the force produced by the contraction of muscles of the posterior trunk is transfered to the anterior end of the body. There, accompanied by local relaxation of the circular muscles, it may cause dilation and enlargement of the burrow. Secondly, the work done in any movement may be derived from the tonus of the whole body wall, involving but a small strain on any muscle. One disadvantage of a single cavity, as pointed out by Batham and Pantin[11] in respect of sea anemones, is that the muscle system of the entire animal is involved in each act. While this is undoubtedly true, it is minimized if the volume of the body cavity is large in comparison to the change in shape, e.g. eversion of the proboscis of *Arenicola* in comparison to the trunk coelom. Additionally the lugworm retains the septate condition in the tail so that the function of defaecation attributed to this region is effectively separated from locomotion.

The production of a high pressure pulse in the coelom involves tension being developed in the trunk musculature. Determinations of muscle tension, for worms of mean diameter 1.2 cm, using the formulae given on p. 10 with circular muscles of average thickness 0.02 cm, give values for these muscles of 0.6 N cm^{-2} at resting pressure and 30 N cm^{-2} at high pressure pulses.[110] The latter value is comparable to values obtained for muscle-strip preparations made from the body wall. Peak pressure is only sustained for about 2 s, however, and the power stroke is undoubtedly the contraction of the relatively massive longitudinal muscles. Chapman[22] has observed that in the earthworm the longitudinal muscles can develop a pressure in the coelomic fluid of ten times that of the circular and suggests that the longitudinals never exert their maximum tension in a free unconstrained worm. In *Arenicola*, also, the longitudinal muscles contract maximally only when a high proportion of the trunk is supported by the burrow. The pressure must be supported by the circular muscles over any part of the trunk outside the burrow, and indeed, some increase in diameter of this region has been observed.

For theoretical consideration of their locomotion, worms have commonly been treated as thin-walled, liquid-filled cylinders, but the observations made by Seymour[99] suggests that the rigidity of their body walls affects the force exerted on the environment. This

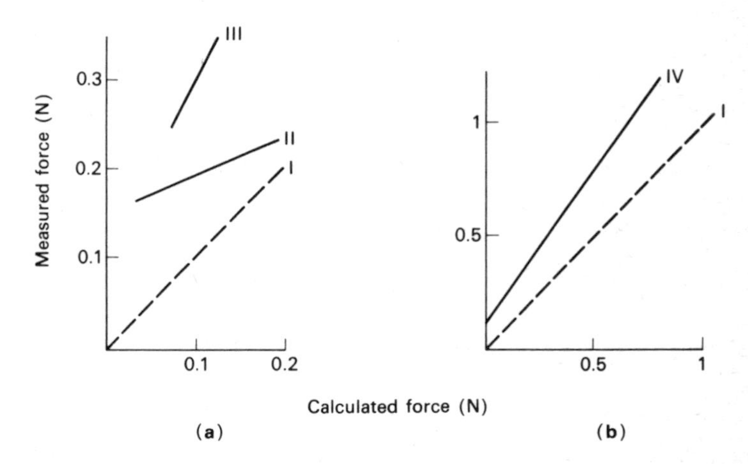

Fig. 3.20 Graphs illustrating the relationship between the measured thrust against a movable bridge exerted by *Lumbricus* (**a**) and *Arenicola* (**b**) and a force calculated as the product of internal pressure and area of worm in contact with bridge. Line I in both graphs represents measured forces equalling calculated forces; in A lines II and III are for worms of 5 and 6.4 g weight respectively; in B line IV is for *Arenicola* of 9.7 and 12.4 g. (after Seymour[99])

was investigated by simultaneous measurements of internal pressure (P) and force exerted whilst the worm was passing beneath a narrow perspex bridge. The bridge was arranged so that the force of displacement of the bridge was recorded by a transducer and the area (A) of the worm's surface in contact with the perspex was recorded on synchronous cine film. A calculated force value ($P \times A$) could then be compared with the force recorded by the transducer. In a worm in which the wall has no inherent rigidity the measured force recorded by transducer would equal the calculated force (Fig. 3.20). In *Arenicola* and *Lumbricus* the measured force exceeds the calculated force, the excess being developed by the body wall musculature. Excess lifting forces for *Lumbricus* and *Arenicola* respectively are 145 and 85% greater than the calculated forces. For the earthworm, measured force and area of contact vary inversely, for the application of increased force results in an increase in the curvature of the worm. Complete intersegmental septa are present in *Lumbricus* and muscle fibres within them are kept taut to resist bulging under the differential pressure of locomotion.[82] Such tension aids the development of a lifting force by restricting lateral spread with circular shortening in a bridge situation or in a crevice in the ground. Additionally, the thicker, and therefore shorter, a region of the earthworm is, the more septa are present and the greater their effect.

In contrast, where the trunk is without septa, as in *Arenicola*, the greater the force exerted against the bridge the greater the area over which it is exerted. This means that the body wall remains more flattened and with the absence of septal musculature does not attain the stiffness of that of *Lumbricus*. *Arenicola* is far less able to develop the unidirectional lifting force required by *Lumbricus* in a crevice situation; but it is ideally suited for the all-round enlargement of a more homogenous substrate where restriction of lateral body-wall spread would be inappropriate.

CONTROL OF BURROWING

Although burial requires the co-ordination of much of the body musculature, little has been written concerning the nervous control of burrowing. There are two reasons for this: firstly, the limited number of animals which will burrow under suitable experimental conditions; secondly, the difficulties inherent in making nerve recordings from an animal whilst moving, often somewhat jerkily. Recent workers have elucidated the functional anatomy of the nervous system of various burrowing animals. Olivo,[83] for instance, has made preparations of the isolated foot, pedal ganglia and cerebro-pedal connectives of a razor clam, *Ensis directus*, and has demonstrated a withdrawal reflex and shown that electrical stimulation of the connectives produces co-ordinated probing movements. These experiments imply that the pedal ganglia contain motor centres which generate reflex movements of the foot, but no one has yet been able to relate the series of body movements carried out during burrowing at the level of nerve responses. An alternative approach to this problem is to investigate the behavioural responses of animals whilst burrowing and the factors which interfere with these, as Seymour[99] has indicated for *Arenicola*.

The West Indian surf clam, *Donax denticulatus*, unlike most bivalves, readily responds to stimulation by activity, and, although small in size, has proved to be a most suitable experiental animal for observations on the co-ordination of locomotion.[120] It burrows in a similar manner to other infaunal bivalves, progressing rapidly (digging period about 4 s) with a rate of penetration of the substrate of 0.4 cm s^{-1} compared with 0.06 cm s^{-1} for the British species *D. vittatus*. *D. denticulatus* occurs in the saturated wash zone of sandy beaches. This habitat is subjected to severe disturbance by waves, so that survival requires the ability to burrow rapidly and repeatedly and to move up and down the beach with the tide. During experiments it was handled by a thread attached to the postero-dorsal region of the valves and this did not impede movement. Either in the laboratory or on the beach the clam could be drawn out of the sand by gently pulling the thread, when the foot would be protracted and was ready to commence probing. Placed on a hard substrate, the foot would make searching movements before the shell closed. The foot would only commence probing if sand grains were present on both sides. If the clam was lifted by the thread and moved with a rotary motion, the valves would open and the foot would be extended whether immersed or in air. Burial commenced immediately if the animal was laid on sand, and disturbance or tactile stimulation of the foot caused a cessation of burrowing only during the first part of the digging period. During the rapid sequence of digging cycles no response was observed. Indeed, when tethered by the thread half buried and apparently quiescent, the response of *Donax* to vibration or movement of the adjacent sand was immediately to attempt to dig more deeply. From observations of this kind during the entire digging period it proved possible to summarize the different phases of burrowing activity diagrammatically, together with some of the factors controlling it in *D. denticulatus* (Fig. 3.21). The early part of

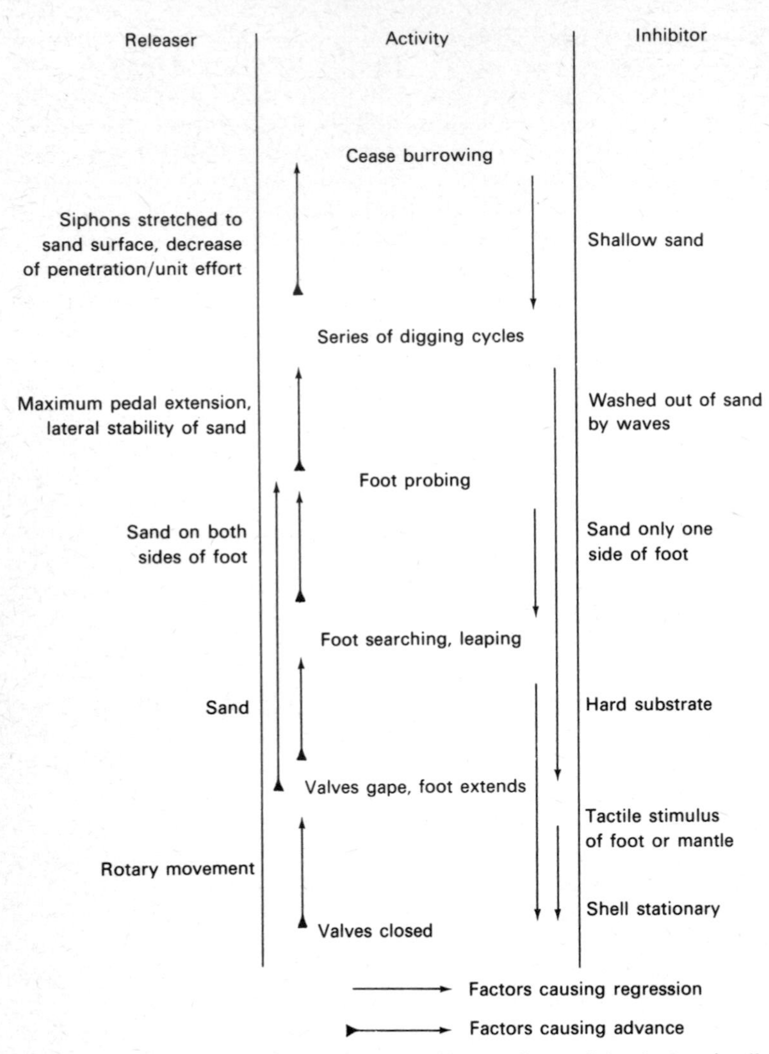

Fig. 3.21 Diagram representing phases of burrowing activity during the digging period of *D. denticulatus* and some factors causing advance or regression of this sequence. (from Trueman[120])

burrowing is essentially a selection mechanism eliminating burrowing attempts on unsuitable substrates; the conclusion of digging is probably related to two factors, the stretching of the siphons to the sand surface and the increase in resistance of the substrate to penetration so that maximum effort is required for minimal progression.

There is a lack of reaction to stimulation during the series of digging cycles of the second part of the digging period. Recognition of this led to further investigation of the factors affecting digging cycles. Experiments made with the valves wedged open to constant gape, or removal of siphons, ligaments or cardinal hinge teeth, made little difference to the burrowing process. They show that the digging cycle phases marked by siphonal closure, adduction and pedal retraction are not affected by external interference, which suggests that these stages of the cycle are programmed within the nervous system and are not dependent on peripheral feedback from proprioceptors located in hinge teeth for example, and affected by movement such as valve adduction. The time spent in probing and pedal retraction does, however, vary with depth of burial and this suggests that stages iv–vi of the digging cycle are controlled by the responses of proprioceptors in the pedal musculature. The use of

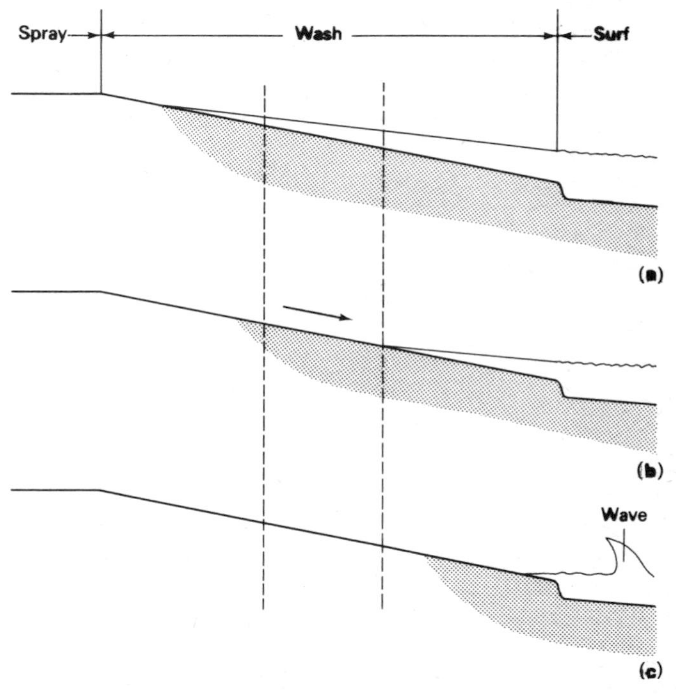

Fig. 3.22 Profile of sandy beach, with the zones marked above, on which *D. denticulatus* can burrow only between the broken lines. In (**a**) a wave has run up the wash zone saturating the sand; (**b**) water running off the beach (arrow) leaving unsaturated sand above and a saturated zone below; (**c**) water almost completely off the wash zone just before the arrival of the next wave. (from Trueman[120])

the foot to sense the state of the sand during digging (Fig. 3.21), and the role of the pedal ganglia in the co-ordination of probing,[83] further emphasize the importance of the foot in the control of locomotion of bivalves.

MIGRATORY BEHAVIOUR OF *DONAX*

Tropical species of *Donax*, for example *D. incarnatus* in India,[8] *D. variabilis* in N. America,[120] are unusual amongst bivalves in that they migrate up and down the shore; in this respect they are one of the more sophisticated and interesting of burrowing animals. They remain in a narrowly defined part of the wash zone (Fig. 3.22) which moves up or down the beach with the state of the tide. Advantages of this ability to maintain position relative to the tide include the possibilities (i) that within this particular zone amounts of suspended organic material may be maximal; and (ii) that the disturbance of continual wave action may afford some protection from predation. Beaches in Jamaica where *D. denticulatus* commonly lives have a slope of about 1 in 10 and as the tidal range is narrow the intertidal zone is also narrow. In the wash zone waves run up the slope saturating the sand. The upper part of this zone is completely drained between uprushes while in the lower region drainage is incomplete and the sand remains saturated between waves. *Donax* is only able to burrow in sand saturated with water where undisturbed by wave action. This clam cannot burrow into dry sand and is only secure against waves when at least two-thirds of the shell is buried. Sufficient burial to hold fast against waves requires a minimum of about 3 s. High up the beach the sand dries out too quickly for burial. The lower part of the shore would thus be the most favourable for *Donax* to burrow into, except that the interval between the backwash and the next wave breaking is minimal. Thus the limits (Fig. 3.22) between which *Donax* can burrow on the beach are set by the opposing factors of the duration of saturation on the upper shore and the wave frequency on the lower. Burial is limited to a narrow zone which is not static but which changes in location with the tidal cycle. Thus clams may burrow in different but specific parts of the beach according to the state of the tide.

Migratory moves are achieved by relatively minor adaptations of the behavioural patterns of an infaunal bivalve. Two points may be stressed: the high rate of burrowing, and the ability to regain the surface after accidental burial. The speed of burial of *Donax* and littoral tropical genera is unequalled outside the tropics; for example, the digging period of *D. denticulatus* is 4 s whereas that of the British *D. vittatius* is about 60 s (Fig. 3.4).

The movement of *Donax* up and down the beach with each tide constitutes a migratory cycle (Fig. 3.23) achieved by the following series of locomotory events, termed collectively a migratory movement. The clams emerge from the sand by pushing downwards with the foot as valves adduct; they are thus lifted by an incoming wave and carried shorewards, when they drop to the sand and attempt to burrow. If they do not settle within the area of the wash zone

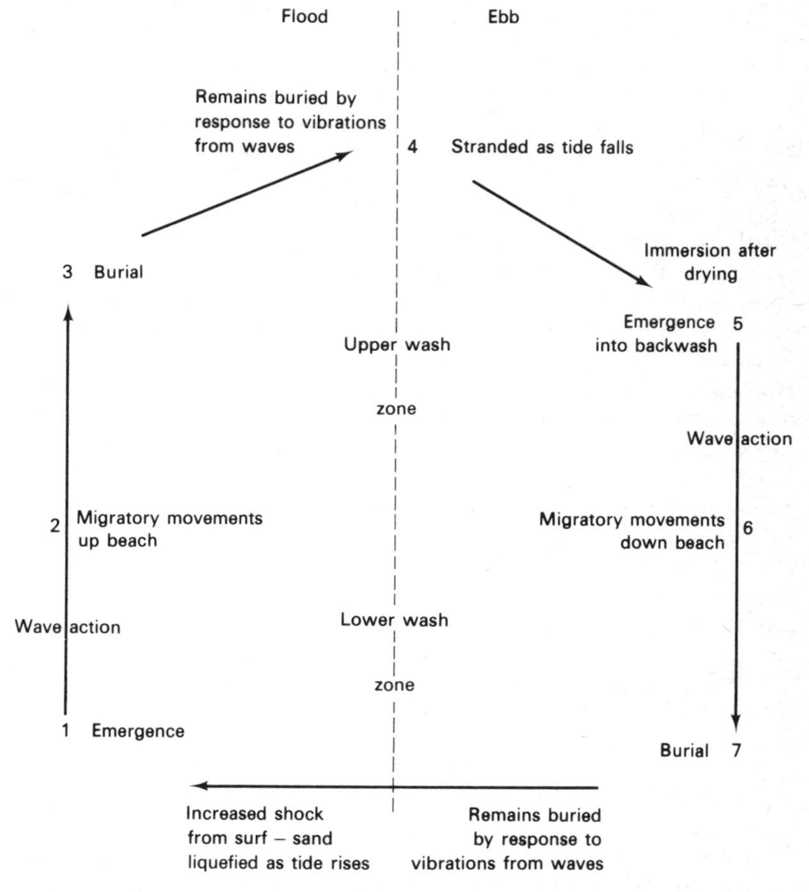

Fig. 3.23 Diagram of the principal events associated with migration of *Donax*. The sequence of actions (1–7) occurring during a migratory cycle is indicated by arrows and the factors influencing these activities are summarized. The broken line represents the change from flood to ebb tide conditions. (after Trueman[120])

designated by the burrowing limits they are washed up and down the beach by successive waves until they are able to burrow successfully. On the rising tide emergence occurs as a response to the increasing fluidity and instability of the sand-water mixture as the buried *Donax* are overtaken by the surf zone. Burial can only be achieved in the wash zone above, so consequently the net movement is up the shore. Migratory movements are repeated on the falling tide when *Donax* emerges from the sand at the top of the beach, because of immersion by a wave following the drying of the sand. The clams now move down the beach for burial on the ebbing tide.

Once the clam has emerged from the sand, movement up or down the beach is caused solely by water currents. *Donax* has little control over this, although it has been suggested that the siphons and foot are used as fins or brakes. Only two locomotory activities are required of the clam to achieve migratory movements, emergence from the sand and rapid burial.

When buried in the wash zone *Donax* is disturbed by each wave breaking. This involves erosion and deposition of sand and it is necessary for a clam to adjust its position relative to the sand surface rapidly so as to be able to feed in the backwash of each wave. This is achieved by actively burrowing and returning to the surface with the passage of each wave.

The energy requirements associated with a migratory life in the wash zone has been calculated for two species of *Donax* (*D. denticulatus* in Jamaica and *D. incarnatus* in India).[10] It consists of two components: (i) movement in or out of the sand to perform a migration; (ii) repeated digging to maintain the feeding position. For the Indian species, a small energy cost of only 87×10^{-2} J (20.8×10^{-2} cal) is estimated to be required daily for migration whilst 26.8 J (6.4 cal) are used daily in maintaining position in the wash zone. Similar figures were obtained for *D. denticulatus*. Thus survival in the wash zone is much more costly of energy than migration. The oxygen consumption of *D. incarnatus* filter feeding at rest is 0.25 mg oxygen/animal/h. This is equivalent to a maintenance requirement of about 83.7 J (20 cal) per day assuming an oxy-calorific equivalent of 14.2 J/mg (3.4 cal/mg). Thus irrespective of migration, survival in the wash zone adds a considerable energy requirement. However, with adaptations of burrowing behaviour and feeding habits, *Donax* is outstandingly successful, for the very instability of the environment tends towards the exclusion of competitors, and its migratory behaviour ensures that it may feed continually whatsoever the state of the tide.

4

Boring into Hard Substrates

When the problems of penetration of hard substrates are considered the habit of boring into rock is surprisingly widespread amongst soft-bodied animals. The advantage of this mode of life is a permanently protected habitat whilst procuring food, either from the burrow, as the sipunculid *Aspidosiphon*,[135] or from the surrounding water as in bivalve molluscs. Animals that bore occur in many invertebrate phyla including some sponges, a flatworm, some sipunculid and polychaete worms, certain echinoid echinoderms, the barnacle *Lithotrya* and a wide range of molluscs, both gastropods and bivalves. That the habit has evolved in such a diversity of groups suggests that it is not primitive and in general animals which bore show specialized adaptations, both morphological and behavioural, for this mode of life. The substrate bored is commonly either soft sedimentary rock or calcareous secretions of animals, notably coral reefs or molluscan shells. Boring occurs commonly in shallow, often intertidal, waters wherever suitable substrates occur in tropical and temperate seas.

It is necessary to define the terms 'boring' and 'burrowing' because of the similarities and differences between these actions, particularly in respect of the Bivalvia. Burrowing may be defined as making an excavation in an unstable substrate, while boring implies piercing a hole in a solid substance, the action often involving rotation about the longitudinal axis of the hole.

THE MECHANISMS OF ROCK BORING

The best examples of boring into solid substances are found in bivalve molluscs but it is useful to review other examples prior to a more detailed examination of the methods used by these animals. Boring is accomplished by mechanical abrasion, by chemical means if the rock is largely calcareous, or by a combination of these. The firmer the substrate the greater is the need for some chemical assistance in boring. Initial penetration cannot be achieved by simple, soft, rasping organs like the proboscis of *Arenicola*. Some hard part of the body must be applied with maximal force, particularly in the absence of chemical solution of the substrate. This requires firm anchorage on the rock surface or more commonly the use of a crevice. Repeated abrasive movements wear out the hardened structures and provision must be made either for their replacement or for continued secretion, as in snail radulae or bivalve shells respectively.

A variety of sponges, notably species of *Cliona*, bore into calcareous rock, coral skeletons and mollusc shells. Larval development is planktonic and after settlement young *C.celata* quickly penetrate oyster shells forming extensive galleries. The method of boring is obscure. Siliceous spicules could assist mechanical abrasion but, in view of the calcareous nature of the substrate, acid secretion would appear more likely. The turbellarian, *Pseudostylachus*, is known to attack oysters, making keyhole-shaped openings into the shell and then eating the contained animal.[135] The means of boring by this flatworm is again unknown but in the absence of jaws or hard structures it must principally involve chemical means. Species of *Polydora* are the best known of polychaete rock borers, always apparently boring into calcareous substrates. These worms form mucous tubes on the surface of oyster shells. Mud adheres to the tubes and the worms move back into crevices and commence to bore. They do so with the body bent into a 'U' shape, the arms being separated by a partition of mud and debris. Boring is at least in part mechanical, possibly by means of enlarged dorsal chaetae on the fifth setigerous segment. Since the substrate is always calcareous, boring may be assisted by some chemical action, although there is no evidence of any appropriate glandular secretory activity.

Descriptive and experimental studies by Carricker[21] on the boring mechanisms of various prosobranch gastropods, for example *Thais*, *Murex* and *Natica*, support the view that both chemical and mechanical mechanisms are involved in boring by snails. These predatory gastropods obtain food by drilling small circular holes through calcareous shells. The radula, a belt of extremely hard teeth, is moved in a rasping action which serves to abrade the shell, whilst an accessory boring organ, located in the sole of the foot in Muricidae and Thaididae, and on the proboscis of Naticidae, secretes a substance which dissolves or softens the shell. Observations on the boring of the whelk *Urosalpinx* into oyster shells show that during the early stages the snail rasps the periostracum from the shell surface and explores the boring site with the foot and proboscis tip. When active boring begins the snail rasps with the radula for a few minutes, retracts the proboscis, and brings the accessory boring organ over the cavity. This remains closely attached for periods of up to nearly an hour, when rasping is continued. The secretion of the boring organ softens and loosens the surface crystals of calcite, so enabling the radula to drill a hole readily. Members of certain families, e.g. Cassidae, Tritonidae, apparently secrete acid to bore into echinoderm prey but the shell-softening secretion of *Urosalpinx* has a neutral reaction and has not yet been identified. The boring mechanisms of various predatory gastropods is reviewed by Owen.[84]

BORING ACTIVITIES OF BIVALVES

Introduction

The literature on molluscs abounds with descriptions of the boring mechanisms of bivalve species and many theories of the mechanism of boring have been put forward. This is hardly surprising because most rock borers are difficult to observe *in situ*, boring activity takes place spasmodically rather than continuously and most species fail to establish a new boring after being forcibly removed. Theories of boring based on visual observations and deductions from morphology have allowed the development of a fairly clear picture of boring habits in the Bivalvia, but the information has remained largely qualitative and descriptive until recently, when Ansell and Nair[7,81] investigated boring using experimental techniques similar to those employed in investigations of burrowing.

Rock boring

The rock boring habit has been adopted independently by no less than seven superfamilies of the Bivalvia, namely, the Mytilacea, Veneracea, Cardiacea (Tridacnidae), Gastrochaenacea, Saxicavacea, Myacea and Adesmacea. Only in some members of the Mytilacea is boring undoubtedly assisted by chemical means; in all others mechanical abrasion by the shell is now considered to be the sole boring device. While animals using chemical means are restricted to calcareous substrates, the majority of bivalves can exploit a much wider range of rocks. The rock boring habit has been arrived at by two main routes: (a) from a primitive infaunal habit by morphological adaptation to burrowing in progressively stiffer muds, as by members of the Myacea (*Platyodon*) and Adesmacea (*Zirphaea*); and (b) from animals attached byssally to a hard substrate where contraction of pedal retractor muscles leads to abrasion of the rock.[135] The latter is notably developed in members of the Mytilidae, e.g. *Botula*[134] (Fig. 4.1) and in the Saxicavacea, e.g. *Hiatella*.

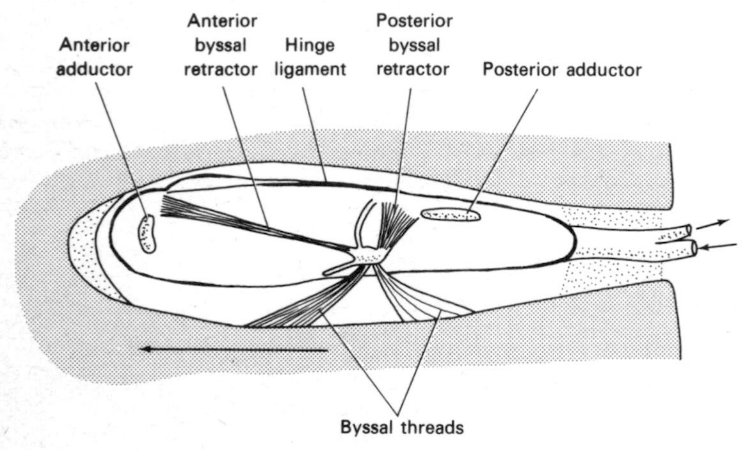

Fig. 4.1 Diagram of *Botula falcata in situ* in boring. Main abrasive action (large arrow) by contraction of posterior byssal retractor muscles. Inhalant and exhalant water currents indicated, rock stippled. (after Yonge[134])

Petricola pholadiformis (Veneracea) shows relatively little adaptation, either of shell form or behaviour, to the boring habit. It inhabits only soft rocks and is able to burrow in sand.[6,7] The animal lies in its substrate obliquely to the surface with the siphons extending through a plug of loosely compacted material that has been removed from the burrow (Fig. 4.2).[31] When removed from sand, clay or chalk and placed on the surface of sand, *Petricola* burrows using the same type of movements as other bivalves (Chapter 3). Initial penetration is carried out somewhat slowly and is followed by a long series of digging cycles. During the latter part of the digging period there occurs a secondary cycle in which a high pressure pulse is developed in the mantle cavity. This is particularly

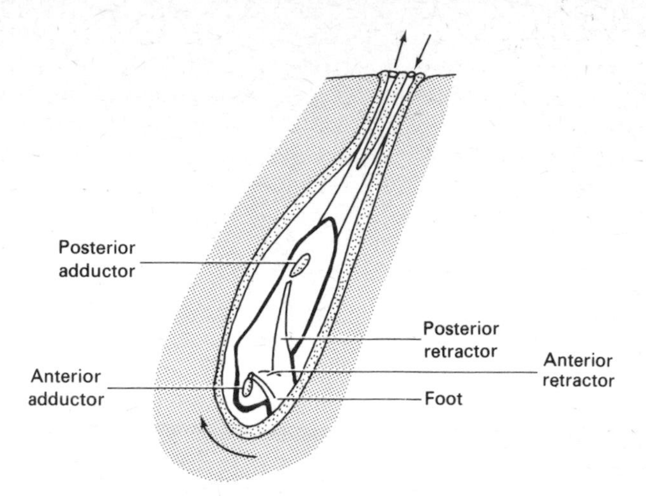

Fig. 4.2 Diagram of *Petricola pholadiformis* in burrow, main abrasive action occurring when posterior pedal retractor contracts drawing anterior of shell upwards (large arrow). Inhalant and exhalant water currents indicated, substrate stippled, coarse stipple representing sludge. (after Duval[31])

well shown during burrowing by *Venerupis* (Fig. 4.3a), where a pressure pulse occurs at adduction during the normal digging cycle; this is associated with siphonal closure and movement into the sand, as indicated by the increase in tension on a force transducer. This is followed by a second pulse of greater pressure, again associated with siphonal closure but also with siphonal retraction and probably pedal retraction. The function of this secondary cycle is to force the valves apart to form a new penetration anchor whenever the energy stored in the ligament is too little to accomplish this. As bivalves burrow so an increasing valve surface area presses against the substrate and the force required to open the valves becomes larger. In *Petricola* the ligament requires supplementation whenever the shell is more than half buried.

The action of *P.pholadiformis* boring into chalk or clay is exactly similar to the burrowing action already described, except that initial penetration is not possible for adult specimens in these substrates. Boring only begins when these animals are placed in artificial burrows which afford them some support. The secondary cycle of siphonal withdrawal also occurs (Fig. 4.3b) and when boring in clay it is apparent that the high pressure serves to consolidate the burrow walls. Excavation of a boring takes place by the foot being extended and dilated to rest against the lower surface over a broad area, so as

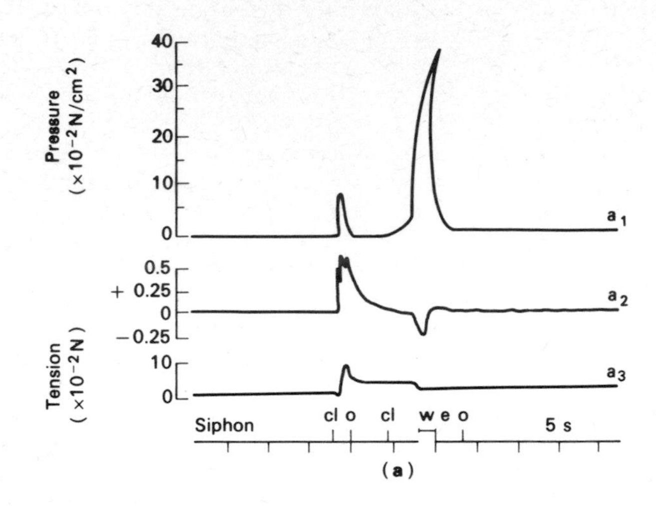

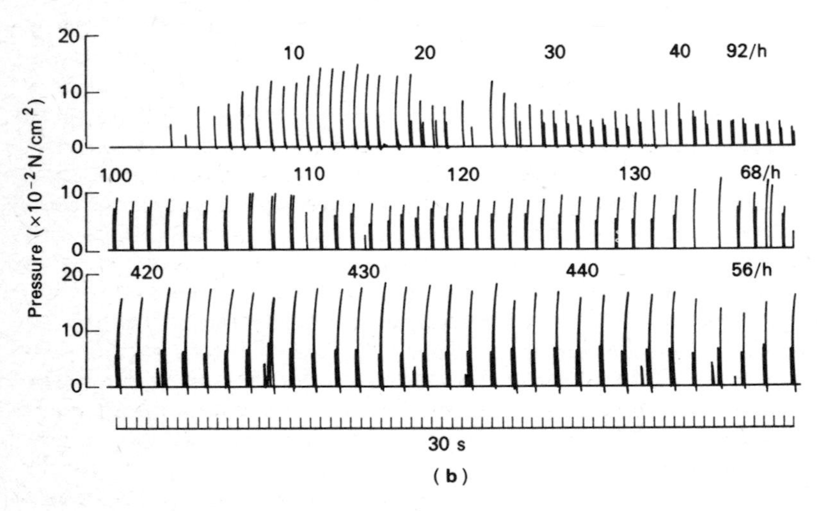

Fig. 4.3 Extracts from recordings of (**a**) *Venerupis* burrowing in sand; (**b**) *Petricloa* boring into London clay. In (**a**) simultaneous recordings of one digging cycle are shown, namely pressure in the mantle cavity (a_1, $\times 10^{-2}$ N cm^{-2}), pressure recorded in adjacent sand (a_2), and tension ($\times 10^{-2}$ N) exerted, measured by a thread from the shell to a force transducer (a_3). Siphonal closure (cl), opening (o) and withdrawal (w) and extension (e) are marked by visual observation. In (**b**) the mantle cavity pressure ($\times 10^{-2}$ N cm^{-2}) is recorded during a long sequence of cycles (numbered), note pressure pulse of secondary cycle of siphonal withdrawal and extension commencing at cycle 19. (from Ansell[6])

to act as a fulcrum about which the valves are rocked in a dorsal direction, principally by means of the powerful posterior retractor muscles (Fig. 4.2). The movements are those of a typical bivalve digging cycle and are repeated at regular intervals for long periods when reforming a burrow (Fig. 4.3b). During normal life, however, the pattern of activity is of long periods with the siphons fully extended whilst a feeding current is maintained, interspersed with short periods of boring. This is probably typical of the normal behaviour of all boring animals where the burrow needs to be enlarged only sufficiently to accommodate the grov ng animal.

The similarity of the mechanism of burrowing and boring in *P.pholadiformis* suggests the importance of pre-adaptations associated with the burrowing habit in the evolution of boring in the Petricolidae. However, major movements of the digging cycle are identical with those used in surface locomotion by other members of this genus and these observations do not contribute firm evidence concerning the origin of boring in the Bivalvia.

Perhaps the most surprising thing about boring in *Petricola* is the development of secondary pulses of pressure in the mantle cavity to force the valves open. This pulse has a duration of about 5 s even although the ventral mantle margins are free and the mantle cavity open ventrally. In *Mya*, a deep burrower, the functioning of the mantle cavity as a hydraulic organ is more advanced, for extensive cross fusion of the mantle lobes has taken place and the mantle cavity opens only by the inhalant and exhalant siphons and the pedal aperture (Fig. 4.4). The hinge ligament is also modified to enable the valves not only to open and close about an anterior-posterior axis, as in other bivalves, but also to rock about a dorso-ventral axis. This motion allows the muscular siphons to be withdrawn between the valves posteriorly as they are pulled together anteriorly by the anterior adductor. More importantly, motion about the dorso-ventral axis, together with the enclosure of the mantle cavity, allows the water in the mantle cavity to function as the fluid of a system whereby the adductors antagonize the siphonal retractor muscles. Provided the siphons and pedal aperture are closed, the mantle cavity remains at constant volume and contraction of the adductors produces a pressure pulse and extension of the retractors. Conversely, the valves can be forced open by means of siphonal retraction in a similar but more effective manner to that of *Petricola*.[112] In *Mya* pressure pulses of up to 15 s duration may reach an amplitude of 1 N cm^{-2}, although 0.4 N cm^{-2} has been commonly recorded. These pressures may be compared with a duration of about 5 s and amplitude of 0.1–0.2 N cm^{-2} in *Petricola*.[7] The ability to

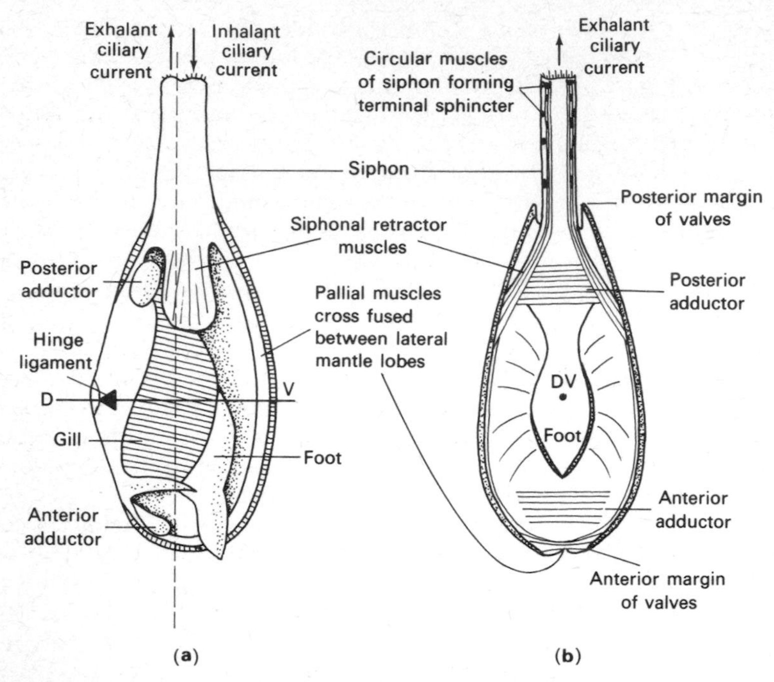

Fig. 4.4 Diagrams of *Mya arenaria* showing the extent of the mantle cavity and the location of the pallial musculature. (**a**) Specimen with left mantle lobe and valve removed, with the foot protruding through the pedal aperture and with the position of the dorso-ventral axis (DV) indicated. (**b**) Represents the dorsal part of *Mya* cut in longitudinal section along the plane indicated by the broken line in (a). Gills and labial palps are omitted to show the extent of the mantle cavity clearly. This section illustrates the antagonistic arrangement of the adductor and siphonal retractor muscles. (after Trueman[112])

generate greater forces to thrust the valves open is clearly associated with ventral mantle fusion and the restriction of outflow from the mantle cavity. Such forces are necessary in burrowing into stiffer muds and indeed *Platyodon*, a close relative of *Mya*, has developed this process so as to bore into relatively soft mudstones.[135] This bivalve bores directly into rock, without twisting, and as a result the boring is not rounded in cross-section but has dorsal and ventral ridges in the region occupied by the shell. Boring is achieved by the pressure in the mantle cavity forcing the valves open, and by the rocking of the valves about a dorso-ventral axis.

Although the boring habit has evolved separately some seven times in the Bivalvia, in no group is it so successful or widespread as in the Adesmacea, where it appears to have been derived from a primitive infaunal burrowing habit in stiffer and stiffer muds in a manner similar to *Platyodon*. All members of the Adesmacea exhibit considerable modifications for boring, but those of the Pholadidae, for example *Zirphaea crispata*, which bore into rock are less specialized than members of the Teredinidae, for example *Teredo navalis*, which bore into wood.

Purchon (1955) described the important anatomical features of *Zirphaea* in relation to rock-boring and these must be considered briefly. Characteristics of the shell include the absence of a ligament, the emargination of the shell anteriorly to provide a wide pedal gape, the sculpture on its anterior outer surface, and the reflexion of the shell in front of the hinge so that part of the anterior adductor muscle is attached dorsally to the hinge axis (Figs. 4.5 and 4.6). The siphons, which are long and fused throughout their length, have

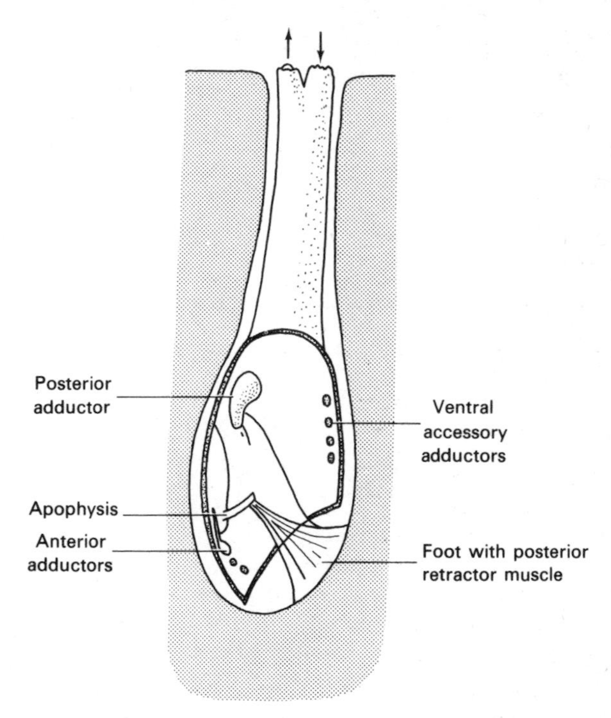

Fig. 4.5 Sectional diagram of *Zirphaea crispata* viewed laterally in boring, main abrasive action produced by contraction of posterior adductor muscles. Inhalant and exhalant water currents indicated by arrows, rock shown stippled. (after Nair and Ansell[81])

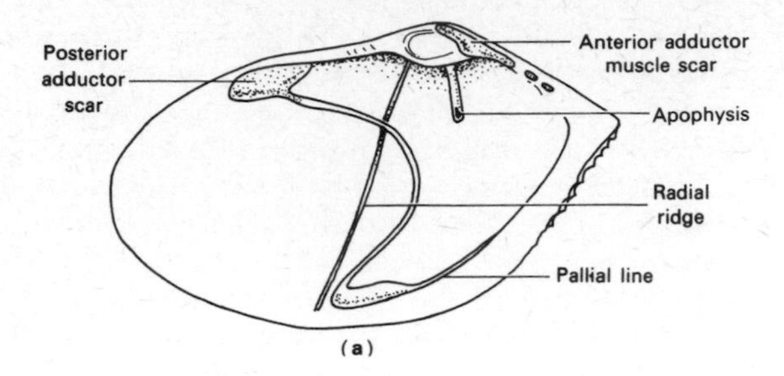

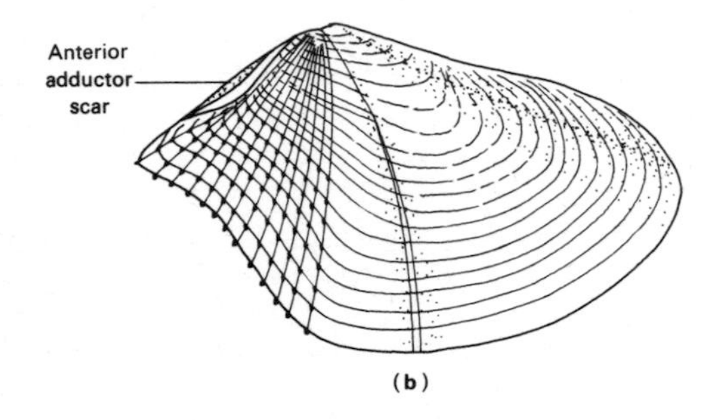

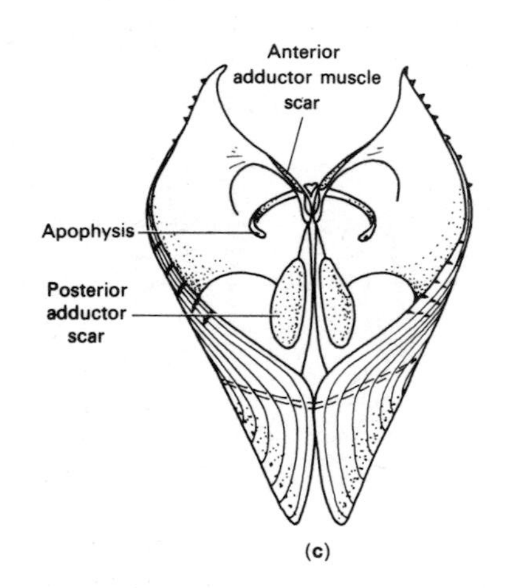

Fig. 4.6 Drawings to show the characteristics of the shell of *Zirphaea crispata*. (**a**) internal view of the left valve; (**b**) external view of the left valve; (**c**) articulated shell valves from an anteroventral aspect to show the position of the apophyses and wide pedal gape. (from Nair and Ansell[81])

their retractor muscles inserted on the shell around a deep pallial sinus and the pallial muscles at their extremity have become modified to form an accessory ventral adductor muscle (Fig. 4.5). *Zirphaea* has the normal bivalve anterior and posterior adductor muscles, except that the anterior is divided into several blocks and is inserted above the hinge line. Thus its contraction can cause the valves to gape, so taking over the function normally performed by the ligament. The pedal musculature is much modified from the basic bivalve type and is relatively poorly developed, so that movements of the foot during boring can only generate weak forces. The pedal musculature consists of three groups of muscles. These are the anterior and posterior retractors, limited to the visceral mass, and the pedal retractor muscles, homologous with the posterior pedal retractors of burrowing forms, which are inserted on a pair of internal processes of the shell, the apophyses.

The mechanism of boring in the pholad *Zirphaea crispata* has been extensively studied by Nair and Ansell[81] and their work forms the basis of this account. Specimens removed from their natural habitat and placed in depressions in the surface of blocks of clay would excavate holes at the rate of about 0.5 mm h^{-1}. The sequence of events during boring make up a series of stereotyped movements which are referred to as the boring cycle. These are best understood by reference to figures 4.7 and 4.8 and may be conveniently summarized:

(i) The foot is extended and applied to the wall of the burrow.

(ii) The siphons close.

(iii) The margins of the foot extend as the tip dilates.

(iv) The pedal retractor muscle contracts and, being invested on the apophyses of the valves, draws them together ventrally. The sole of the foot remains pressed to the wall of the burrow and the shell is drawn downwards.

(v) The shell valves rock about a dorso-ventral axis passing through the umbones of the valves and the accessory ventral muscle, by contraction of the posterior adductor so that the anterior margins of the shell diverge. The siphons reopen.

(vi) The shell rocks back on the same axis by contraction of the anterior adductor restoring the orientation of the valves about the dorso-ventral axis.

Following pedal extension and dilation in stages (i) and (iii), the foot is pressed against the wall of the boring and acts as a fixed point on which the shell can move. It is aided in this by a pad of tissue surrounding the anterior adductor which is distended with blood and pressed against the wall of the burrow. The resemblance

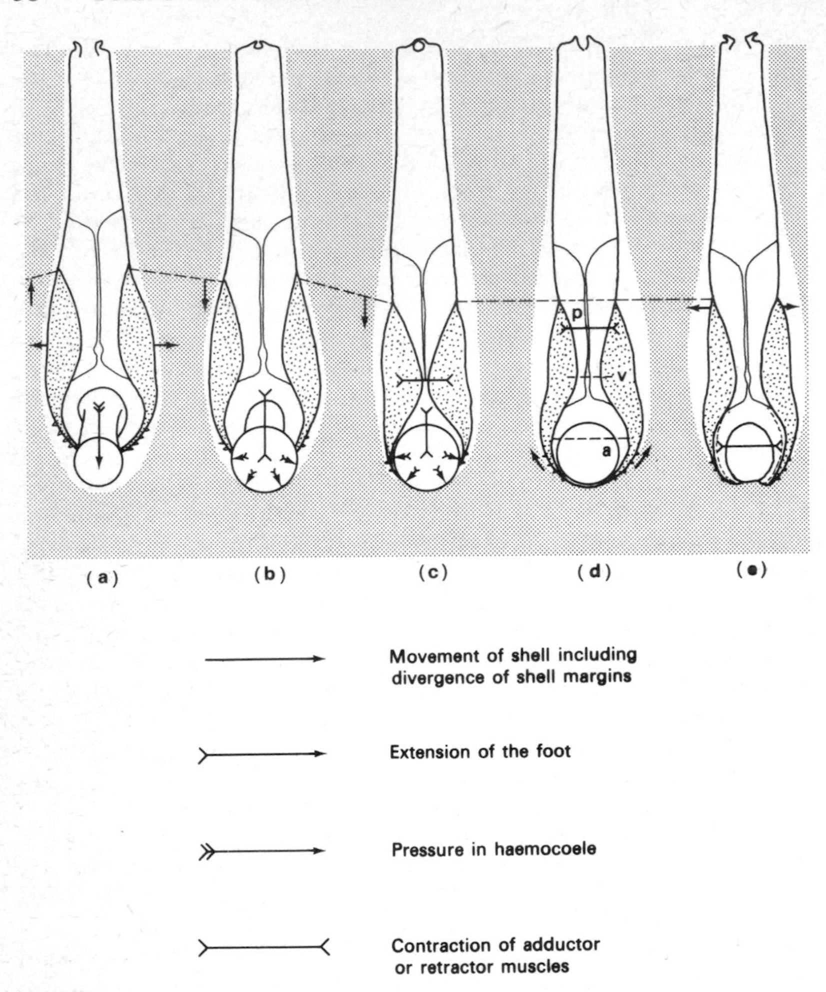

Fig. 4.7 Series of diagrams of *Zirphaea crispata*, viewed ventrally, at different stages of the boring cycle (as in Figure 4.8). (**a**) foot extending, with siphons open (i); (**b**) siphons closed, tip of foot dilated with the margin extended, retraction starting (iii–iv); (**c**) retraction continues, ventral margin of valves drawn together with the action of accessory ventral adductor muscle (iv); (**d**) contraction of posterior adductor muscle, anterior areas of valves abrade the burrow as they diverge (a), siphon re-opening; (**e**) siphons open, contraction of anterior adductor muscle and posterior margins of shell diverge (vi). The lines marked by the letters, a, v, and p, in (**d**) mark the positions at which the anterior, ventral and posterior gape of the shell was measured for the construction of Figure 4.8. (from Nair and Ansell[81])

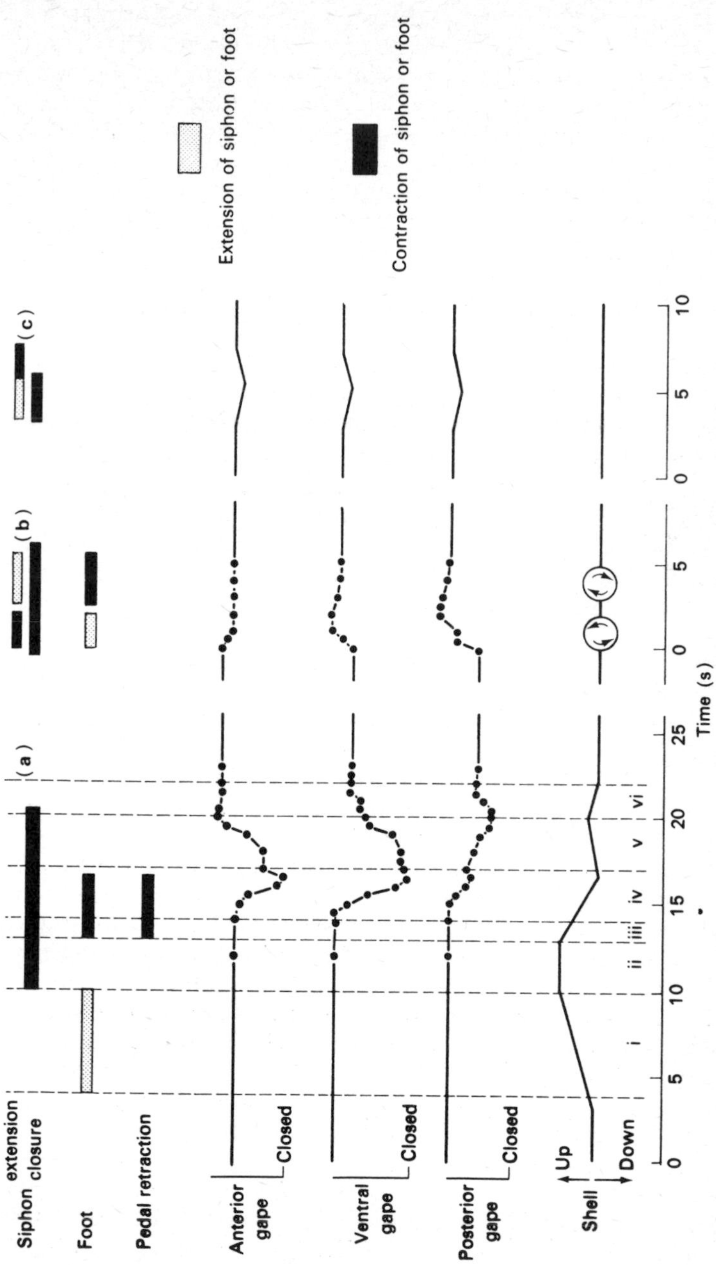

Fig. 4.8 Diagram constructed from film records showing detail of the activity of *Zirphaea crispata* during boring. Movements of the siphons, foot and shell are shown during the boring cycle **(a)**, siphonal retraction and extension **(b)** and pseudofaeces extrusion **(c)**. Anterior, ventral and posterior gape were measured at the points indicated in Figure 4.7 **(d)** by the letters a, v and p respectively, and are taken from a single film record. Vertical and rotatory shell movements are placed in their correct sequence by observation of many cycles. Stages in the boring cycle are indicated, i–vi. (from Nair and Ansell[81])

between boring and burrowing lie in stages (i–iv) of the boring cycle, when the foot forms an anchor and the partially closed shell is drawn down towards the base of the excavation. The function of the foot is thus similar. Adduction of the shell valves during stage (iv) also serves one of the same purposes as in burrowing, namely to increase the pressure in the haemocoel so as to aid pedal anchorage and to reduce the profile of the shell during downward movement. The basic difference between the two activities is that in burrowing, downward movement is the effective stage in penetration but in boring it is but a preliminary phase. In boring the successive contraction of posterior and anterior adductor muscles follows pedal retraction (Fig. 4.9a) and causes abrasion of the boring by rocking the valves. The boring of *Zirphaea* is circular in cross-section which suggests that the shell rotates during the process of boring. This occurs coincidently with siphonal movements, retraction and extension, following each cycle. Anticlockwise movement of between 25 to 30° is followed by a similar clockwise rotation. A second type of rotation results from changes in the position of the foot in the burrow so that the action of the pedal musculature causes the rotation of the shell and indeed the entire animal in relation to the foot. A complete rotation of 360° occurs by this means over about 70 min.

An important aspect is the antagonistic action between the siphonal retractor muscles and the posterior adductor during boring by *Zirphaea*. When the tips of the siphon are closed the pallial water system and haemocoel operate at constant volume so that if the siphons are withdrawn the pressure rises and the valves gape, particularly posteriorly, in a similar manner to that observed in *Mya*. This occurs during boring when the siphon is retracted following the boring cycle (Fig. 4.9a) and pressure is maintained by slight ventral and posterior adduction when the siphons are re-extended. These movements of the shell are accompanied by a lifting and lowering of the shell in the boring and are thought to re-orientate the valves prior to the commencement of the next boring cycle.

The movement of the shell loosens material beneath the bivalve which is collected in the mantle cavity by a ciliary tract around the inner margins of the pedal gape. This material is bound together with mucus and accumulates at the base of the inhalant siphon to be expelled as pseudofaeces by adduction of the valves and the controlled relaxation of the circular muscles of the siphon so as to produce a peristaltic wave travelling from base to siphonal tip (Fig. 4.8). It is during this process that the greatest pressures (Fig. 4.9b) were recorded from the mantle cavity, but even these pressure

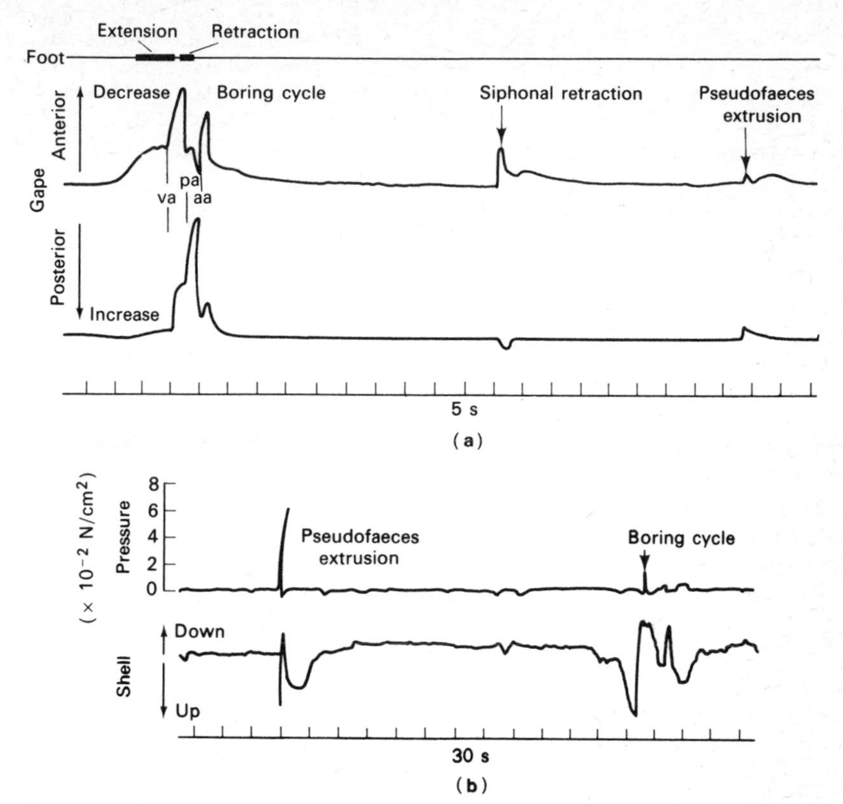

Fig. 4.9 **(a)** Recording showing the effect of the antagonistic action of the adductor muscles on shell movements during the boring cycle, siphonal retraction and extension and pseudofaeces extrusion. *Zirphaea* was fixed horizontally in a dish, immediately after removal from the burrow, and gape recorded by movement transducers with threads attached from the anterior and posterior extremities of the upper valve. Visual observations (upper trace) show extension and retraction of the foot during stages (I) and (III–IV) of the boring cycle. contraction of muscles is indicated. **(b)** Simultaneous recordings of pressure ($\times 10^{-2}$ N cm^{-2}) in the mantle cavity and of vertical movements of the shell of *Zirphaea crispata* buried in clay. Greatest pressures are associated with pseudofaeces extrusion. (from Nair and Ansell[81])

pulses are of low amplitude and justify the conclusion drawn by Nair and Ansell[81] that a high pressure hydraulic system is not a feature of boring in pholads. This is in contrast to the high pressures recorded in some burrowing species, for example *Ensis* and *Mya*, but it can be explained by the manner in which the adductor muscles antagonize each other directly. The umbones of the valves act as a ball and socket joint and the hydraulic system is replaced, at least partially, by the shell acting as a jointed exoskeleton. The full development of this type of movement is reached in the wood boring genus *Teredo*.

Wood boring by *Teredo*

Teredo is perhaps the most specialized mollusc, the mantle tissues being considerably extended into a siphonal tube to form a long wormlike body and the shell being reduced to two small valves (Fig. 4.10). The valves are articulated by dorsal and ventral knobs and

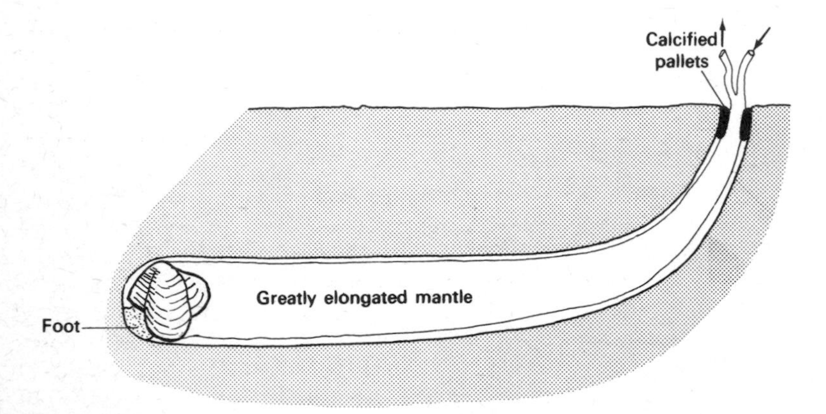

Fig. 4.10 Sectional diagram of *Teredo in situ* in boring. Main abrasive action by posterior adductor muscle contracting, so forcing anterior region of reduced shell outwards. Inhalant and exhalant water currents indicated, animal progressing in direction of grain.

movement only occurs about the dorso-ventral axis (Fig. 4.12e). The shell forms a highly specialized cutting tool. Each valve comprises 3 lobes: the outer surface of the middle and anterior have sharp ridges, which are continually replaced in life, while a powerful adductor muscle is situated between the posterior auricles. The foot is rounded and grips the end of the boring by suction, when the pedal retractors contract, to press the shell against the wood. Contraction of the posterior adductor draws the auricles together and the anterior cutting edge of the valves apart, so rasping away the wood which is to be eaten by *Teredo*. Restoration of the valves is carried out by a much smaller anterior adductor muscle without any abrasive action. After each movement the foot moves around the burrow, rotating both the shell and the location of scraping. Alternation of scraping and rotation proceeds through 180° when the foot reverses. This action produces a boring of circular cross-section. As boring continues the body elongates since the shipworm is attached to the end of the burrow by paddle-like calcareous plates, termed the pallets (Fig. 4. 10). These also serve to close the end of the tube when the siphon is withdrawn. In the mechanical process of boring there is no requirement of a hydraulic system since the muscles operate directly to impart a cutting motion of the valves.

Board[15] has made some further interesting observations. Water from the mantle cavity can be released anteriorly into the boring through the pedal opening and examination of soft wood blocks by radiography shows that the shipworms wet the wood in front of the boring when tunnelling in air-filled wood (Fig. 4.11). The functions of this water are first to lubricate the valves, dispersing the heat generated by friction and removing particles of wood, and secondly to soften the wood. In air-filled wood, water shows a tendency to flow along the grain so that *Teredo* burrowing in this direction would always have adequate water in front, whereas those travelling across the grain may be deficient. Accordingly tunnelling is generally along the grain except when the wood is already saturated with water.

On the basis of analysis of many radiographs and a knowledge of the functional anatomy of *Teredo*, Board suggests that the supply of sea water to the end of the tunnel is controlled by the pressures

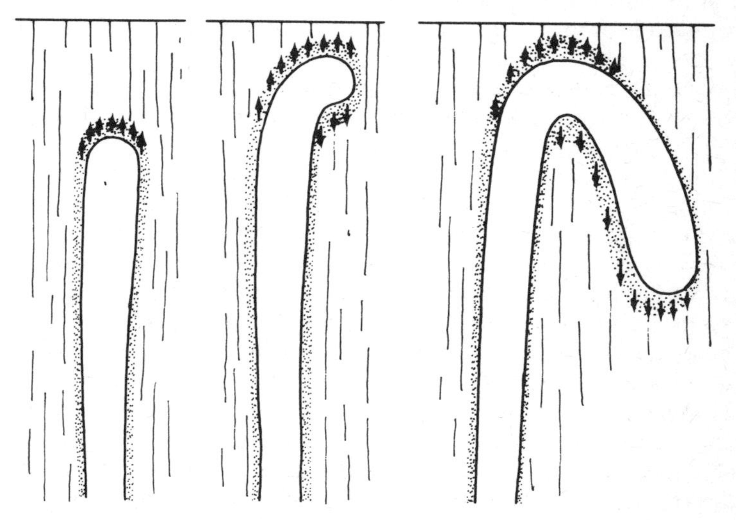

Fig. 4.11 Sections of tunnels of *Teredo* showing wood lubricated by water (stipple), how lubricant moves along the grain (arrows) rather than across it and how the animal avoids the surface of wood. (from Board[15])

generated by the shipworm when boring. Withdrawal of the valves produces a negative pressure which draws water forwards through the pedal opening. This allows lubrication of the valves or partial evacuation of the tunnel when no further progress can be made. A shipworm re-occupies the tunnel in the same manner as when boring but without movement of the valves. Gradual forward progression with a rotary motion, and the probable formation of a water-tight seal with the foot and cephalic hood, allow the shipworm to act as a plunger within the tunnel, forcing water into the wood at the blind end. Cellulolytic bacteria and fungi causing soft rot are found particularly near the surface of water saturated wood. They facilitate initial penetration by juvenile *Teredo* but subsequently prevent their escape from the timber. This is because of their effect in softening the wood which collapses when the animal braces itself for the valves to rasp. *Teredo* is thought to correct this tendency by turning away from the softer wood and accordingly never reaches the surface (Fig. 4.11).

EVOLUTION OF THE BORING HABIT

The habit of rock boring is widespread throughout the invertebrate phyla, although many of the borers are not specifically adapted for this mode of life. In many boring is not obligatory but facultative, as in the sponge *Cliona*, and in *Polydora*, or amongst the bivalves in *Hiatella* and the less specialized species of *Petricola*. All are members of the epifauna with a basic sessile or nestling habit which leads to boring where conditions are suitable. Specialized obligatory borers include certain sipunculid worms, the carnivorous shell borers including the turbellarian *Pseudostylachus*, the prosobranch gastropods such as *Natica*, and many of the bivalve borers. The last-named group, which are the most fully investigated, abrade the rock by lateral and forward movements of the valves.

The boring habit of the highly specialized Adesmacea has clearly been derived from infaunal bivalves and the main line of evolution of the rock and wood boring habits is summarized in figure 4.12. In generalized bivalves, such as *Mercenaria*, the valves can only rotate about the antero-posterior axis of the hinge line, the adductors being antagonized directly by the elasticity of the hinge ligament and indirectly through the hydraulic system of the mantle cavity and haemocoel. In deep burrowing forms, for example *Mya*, the ligament retains its primary function but, being reduced in length, allows movement about the dorso-ventral axis so that the siphons

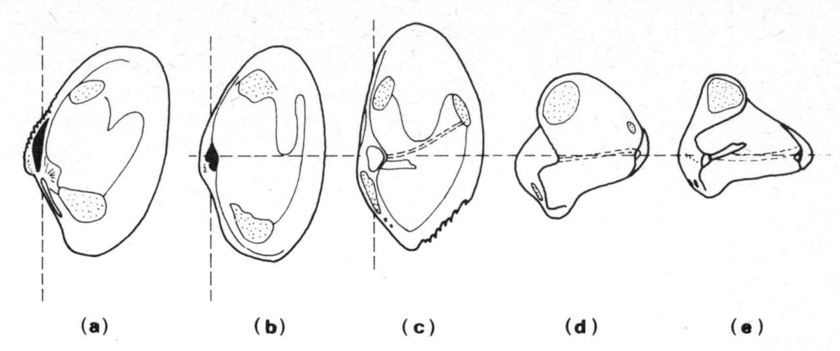

Fig. 4.12 Diagrams illustrating the main line of evolution of the rock- and wood-boring habits of bivalve molluscs. (**a**) *Mercenaria mercenaria*, representative of shallow-burrowing forms; (**b**) *Mya arenaria*, representative deep-burrowing form; (**c**) *Zirphaea crispata*, representative rock-boring pholad; (**d**) the wood-boring from *Xylophaga dorsalis*; (**e**) *Teredo navalis*. The antero-posterior axis through the hinge (**a–c**) and the dorso-ventral axis (**b–e**) are indicated by a broken line. (from Nair and Ansell[81])

may be accommodated on withdrawal. This tendency has been developed further in *Zirphaea* where the ligament is lost. In the Pholadidae, the anterior and posterior adductors effectively antagonize each other about the umbones, the posterior part of the shell affords partial accommodation for the siphons, and antagonism between adductors and siphonal retractors through the hydraulic system is still important. In *Teredo*, the adductors antagonize each other directly across the fulcrum formed of dorsal and ventral articulations; the antagonism between siphons and adductors is no longer important because of the unique posterior extension of the body with the displacement of the siphons from the proximity of the shell valves. Specialization of the shell has here proceeded to such a stage that it has almost lost its protective significance and remains only as a drilling tool, *par excellence*.

In *Tridacna*, where the ligament is not reduced, those that bore grind their way downward by the alternate contraction of anterior and posterior pedal retractors rocking the shell in the longitudinal plane. In the rock boring mytilid, *Botula*, contraction of the posterior byssal retractor muscles causes the anterior end of the shell to bear against the head of the boring whilst the ligament, powerful because of secondary elongation, brings about lateral widening (Fig. 4.1). In the closely related *Lithophaga*, mechanical boring is reduced and a secretion from the mantle glands softens the calcareous rock to which it is confined. As with most other boring organisms we lack information of the precise manner in which the rock is softened.

Protection is the major biological advantage of rock boring, for once penetration has been achieved enlargement of the boring need only take place slowly in step with growth. Only where borers are numerous, particularly on coral reefs, are they major agents of erosion. Reefs are generally maintained, however, by the exceptionally high rate of calcification possessed by hermatypic corals.

5

Crawling and Undulatory Swimming

INTRODUCTION

An important means of locomotion yet to be considered here is the type of swimming in which the body is thrown into waves that pass along the animal, usually from head to tail. Such undulatory swimming occurs in the lower chordates and polychaete worms with axial or fluid skeletons respectively and may be seen to be associated with a segmental arrangement of muscles in both groups. We have already seen in the earthworm the interaction between metameric segmentation and the functioning of a hydrostatic skeleton in respect of burrowing and before considering undulatory swimming in the polychaetes attention must be paid to how these animals crawl by use of parapodia.

CRAWLING

Nereis

The use of parapodia in crawling is a method of locomotion characteristic of such errant polychaetes as *Nereis*, originally described by Gray and summarized in his book on Animal Locomotion[43] and by Clark.[26] The parapodia are used as levers in a similar way to the legs of land animals and in slow crawling provide the entire locomotory force, the muscles of the body wall playing very little part. However, the longitudinal muscles play an important role in fast crawling and swimming and with parapodia represent two distinct effector mechanisms.

To understand the locomotion of *Nereis* brief reference must first be made to its anatomy (Fig. 5.1). Segmentation of muscles and of

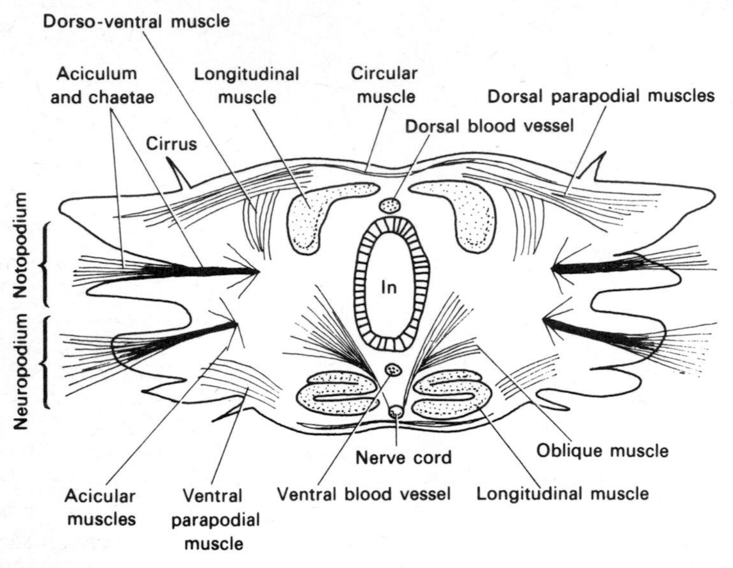

Fig. 5.1 Diagrammatic section of *Nereis* showing arrangement of musculature and parapodia.

their nerve supply from the nerve cord, as in other annelids, facilitates the passage of locomotory waves along the body. Important differences from other polychaetes, however, are the division of a continuous layer of longitudinal muscles into a pair of dorsal and a pair of ventral muscle blocks. By this means the muscles of the two sides of a segment can contract and relax in the opposite phase, enabling the passages of waves which cause lateral undulations of the body. The common arrangement of muscles in cylindrical wormlike bodies is for circular muscle to be opposed to longitudinal muscle but in *Nereis* the circular muscle is relatively weak and incomplete where muscles run into the parapodia. The body wall is thus weaker than that of *Arenicola* (Fig. 1.2b) or the earthworm and some support to the body is undoubtedly given by the oblique muscles. In the absence of circular muscle fibres completely encircling the body, *Nereis* is clearly unsuited for the peristaltic movements so characteristic of other worms. It is the parapodia and lateral movements of the body that are important in *Nereis*. The parapodia are hollow extensions of the body wall formed laterally in each segment and are typically divided into dorsal and ventral lobes called the notopodium and neuropodium (Fig. 5.2). Each lobe is

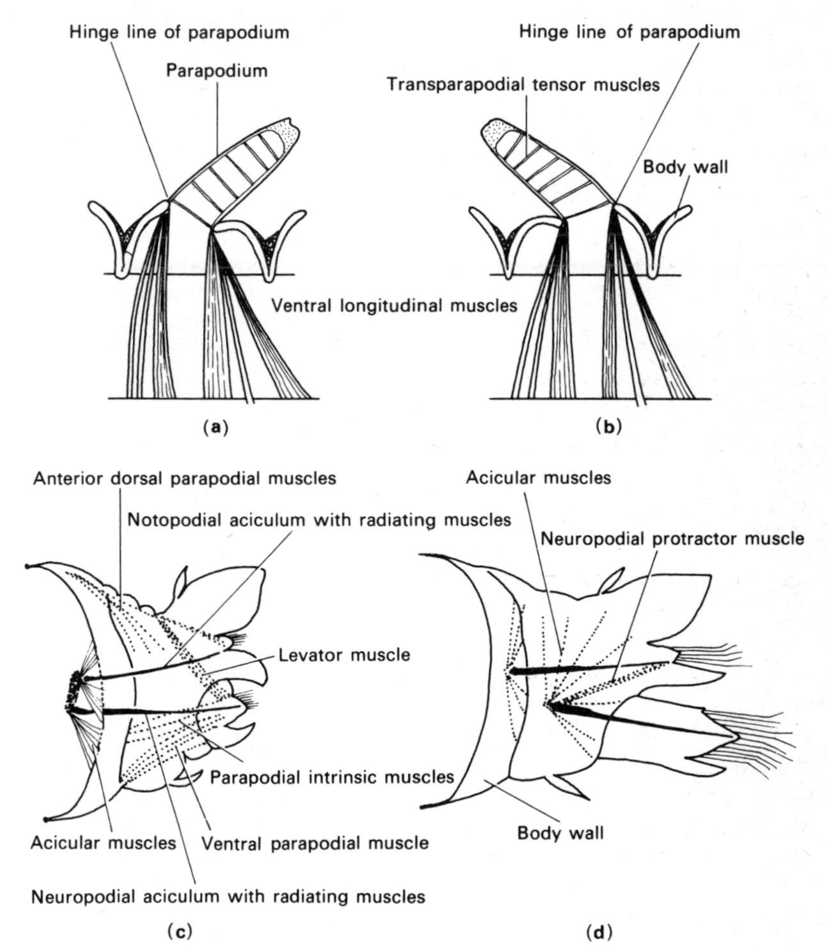

Fig. 5.2 *Nereis diversicolor:* diagrams to show muscle disposition and action. (**a**) Parapodium bent forwards by contraction of anterior parapodial oblique muscles and (**b**) backwards by posterior ones. (**c**) Profile of parapodium retracted during preparatory stroke of slow creeping and (**d**) protracted during power-stroke. Acicula black; contracting muscles shaded as if seen by transparency. (from Mettam[74])

supported by a bundle of bristles or chaetae, strengthened by a supporting aciculum. The coelom extends into each parapodium and parapodial protrusion can be effected by the pressure of the coelomic fluid. Retraction is brought about by muscles whose origin is in the mid-ventral line of the body wall and insertion is dorsally or ventrally onto the parapodium[74] (Fig. 5.2). They are thus termed extrinsic parapodial muscles. There are typically anterior and posterior pairs of these oblique muscles so that they may effect the anterior and posterior movement of the parapodium, the points of attachment to the body wall acting as a hinge. Intrinsic muscles, i.e. those whose origin and insertion is within the parapodium, are largely responsible for protraction and retraction of the chaetal bundles and for changing the shape of the parapodium.

Slow crawling depends entirely on the use of the parapodia as a series of levers. Initially a parapodium is directed forwards, its tip applied to the substrate, and contraction of the flexor muscles exerts a forward thrust on the body. This is the powerstroke, after which it remains inactive, inclined obliquely backwards until it is lifted from the ground and moved forwards in a rapid recovery stroke immediately prior to the next powerstroke. Chaetae are protracted during the back stroke, helping to anchor the tip of the parapodium, whilst during the recovery stroke they are retracted. As one parapodium carries out its powerstroke so the next anterior follows after a brief interval (Fig. 5.3). The phase of movement shown by muscles and parapodium of one side of the body is always half of a cycle out of phase with that of the opposite side. At the commencement of slow crawling the first parapodia to be active are near the head of the animal but the pattern of parapodial movement rapidly spreads posteriorly. During slow forward progression the waves of activity pass over the parapodia on either side of the body alternately and travel towards the head. The worm is thus provided with a great number of *points d'appui* simultaneously and must experience a fairly uniform forward thrust along its body, which is thus in continual motion. The parapodia are, however, being continually accelerated and retarded. Since their mass is small compared to the remainder of the body less energy is likely to be dissipated by this movement than by the starting and stopping of the entire body wall as in peristaltic locomotion. Indeed the situation is somewhat similar to crawling in *Helix* when the snail continually moves forward and only the sole of the foot is alternately moving forwards and remaining stationary as an anchor.

Nereis exhibits a more complex locomotory activity when crawling rapidly, for this additionally involves the longitudinal muscles and as

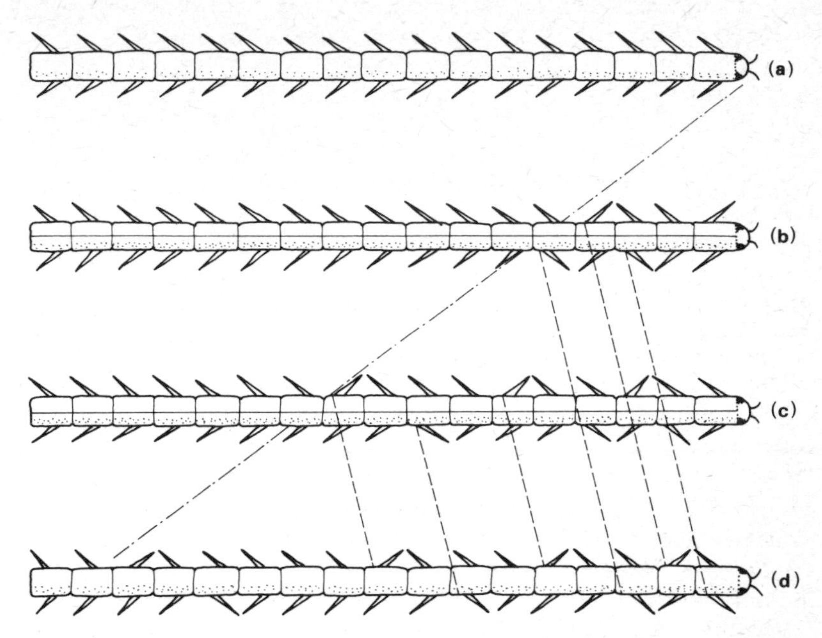

Fig. 5.3 Commencement of slow crawling in *Nereis* (**a–d**); broken line indicates backward spread of locomotory pattern; dotted lines connect successive parapodia about to execute a power stroke. (after Gray[43])

a consequence much more powerful forces are available. Contraction of the longitudinal muscles throws the body into sinusoidal waves passing towards the head as a direct wave (Fig. 5.4). The parapodium on the crest of a wave, where longitudinal muscles are relaxed, will always be stationary in respect of the substrate. Its partner on the opposite side of the segment is in the trough of the wave, moving as the longitudinal muscles contract. During this contraction these muscles exert a thrust on the substrate transmitted by the parapodium on the opposite side of the body. The section of figure 5.4 showing parapodial movements illustrates that as each parapodium remains stationary in respect of the ground, its fellow rotates about it and is moved forwards. As each parapodium is temporarily anchored its underlying longitudinal muscles are relaxed and the adjacent part of the segment is maximally linearly extended. This conforms to the principle of locomotion by a direct wave, when anchorage is always obtained at maximal extension as in the foot of *Helix* (p. 37) or body wall of *Polyphysia* (p. 46). It is the converse of the situation in the earthworm, using retrograde waves,

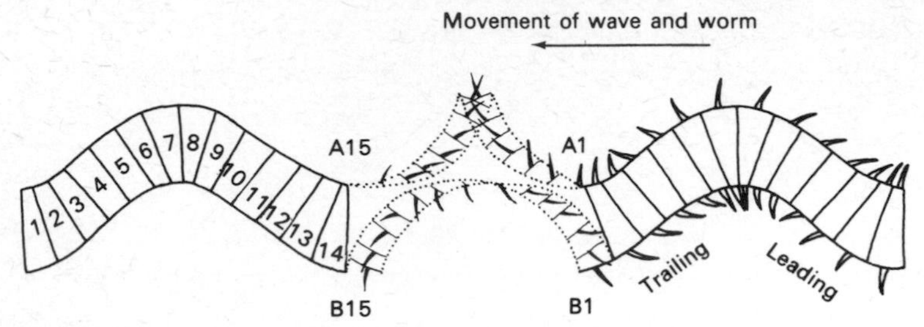

Fig. 5.4 Diagram of *Nereis diversicolor* during fast crawling. Central section shows parapodial movements relative to the ground of a segment at successive positions (1–14) of one complete wave of lateral undulation viewed from above. Positions of successive parapodia at the very crest of the wave are superimposed. There is slight slip relative to the ground. Trailing and leading surfaces of locomotory wave are indicated. (from Mettam[75])

where the anchored segment is always maximally longitudinally contracted. It should also be noted that the locomotory waves of an earthworm extend transversely across the animal and are thus monotaxic, whereas those of *Nereis* are ditaxic and out of phase.

During rapid crawling there are thus two sets of forces contributing to locomotion, from the parapodial muscles and from the longitudinals. The former are only weak and can move the worm only slowly, while much more powerful forces are derived from the longitudinals. The parapodia and their muscles thus appear relatively ineffective as appendages, but their occurrence is of great evolutionary importance, for they represent a major departure from the type of locomotion by peristalsis seen in other worms. It foreshadows the use of legs, rather than body wall muscles, a use which reaches its zenith in the arthropods.

Nephtys

The mechanism of *Nereis* is probably typical of many other polychaetes but there are undoubtedly variations of this pattern of locomotion such as occurs in *Nephtys*.[27] *Nephtys* is an actively burrowing errant polychaete and its method of burrowing is essentially to make a cavity in the sand by violent extrusion of its proboscis. The anterior segments (5–10) are narrower than the proboscis but as an undulating locomotory wave passes, the parapodia on the crest are carried laterally sufficiently far to make contact with the walls of the burrow and perform their powerstroke. The succeeding segments (15–45) also become involved in this activity but since they are almost as wide as the burrow the locomotory wave is here reduced in amplitude. Behind segment 50 the segments taper and are being passively dragged forward without making any positive contribution to the total locomotory forces.

Sabella

Another modification is found in *Sabella*[26] in respect of the worm crawling along its tube. This worm fits its tube tightly, in contrast to *Nephtys*, so that locomotory waves produced by longitudinal muscles are precluded. During crawling a parapodium is extended so that its tip meets the wall of the tube and is swung forwards or back according to the direction of movement. This activity of *Sabella* is likened to 'poling' a barge.

Aphrodite

Aphrodite is a worm with a short wide body which burrows just beneath the surface of sand or mud. Its locomotion and functional design has been recently described by Mettam[75] who divides the segmental muscles into two groups: those controlling body shape and those involved in locomotion. The longitudinal muscles are much less in evidence than in *Nereis*, the dorsal pair being particularly thin, and they have no propulsive role in locomotion. The stronger contraction of the ventral longitudinals causes the worm to curl up like a hedgehog, presenting spines to any aggressor. The circular muscles are extremely reduced, and restoration of shape following contraction of the longitudinals is carried out by pairs of diagonal muscles, which form a muscular lattice.

In locomotion, movement of the parapodium is effected entirely by the acicular muscles and intrinsic parapodial muscles. It can be regarded as a development of the nereid slow crawling mechanism, although the parapodia of *Aphrodite* are much modified. *Aphrodite* employs a fast stepping pattern (Fig. 5.5), the movement of each parapodium invoking extension and retraction of the neuropodium with coordination of the chaetal movements. The downward and backward propulsive stroke raises the worm clear of a hard substrate. The notopodium is stabilized, its long chaetae being used in defence. In association with the longitudinal muscles the body of *Nereis* is long and thin to allow undulatory waves; in *Aphrodite* the body is short and thick, unsuited to undulatory propulsion but suitable for the development of powerful limbs. Manton[69] has emphasized that, with such limbs, body undulations only waste

Fig. 5.5 Diagrams of *Aphrodite*, viewed from below, crawling during one half of a parapodial cycle. Drawn from cine film at 1 s intervals. (from Mettam[75])

energy and that terrestrial arthropods have evolved a variety of means to reduce the tendency to undulate. The short body of *Aphrodite* may be associated with increased parapodial thrust and prevents unwanted undulations; its locomotion has clearly evolved from slow stepping such as occurs in *Nereis*. *Aphrodite* represents an extreme polychaete adaptation to powerful limb movements and in some ways foreshadows the development of locomotion using limbs in the arthropods. Indeed, Mettam[75] observes the similarity of locomotory pattern in *Aphrodite* with that of a terrestrial arthropod such as *Lithobius*.[43]

UNDULATORY SWIMMING

Introduction

Most animals that have elongate, narrow bodies swim by means of undulations passing along the body; as a result, backthrust is exerted against the adjacent water so as to provide motive force. The fluid yields as the body moves, so that the forces available for locomotion are much less than if the animal were in contact with immovable objects. This has been clearly demonstrated by Gray[43] in respect of the movement of an eel across a board both with and without vertically arranged pegs. (Further reference should be made to this author's book on Animal Locomotion for an account of swimming in fish and a theoretical treatment of undulatory propulsion in invertebrates, for only a simplified explanation of undulatory propulsion is presented here.) However, provided the animal is reasonably well streamlined, frictional forces resisting motion during swimming are small.

Both inertial and viscosity stresses must be taken into account when considering movement through a fluid. With large animals, such as fish, the stresses due to viscosity are minimal but the animal relies on the inertia of the water to generate a forward thrust. The ratio of inertial to viscous forces, termed the Reynolds number (R), is an expression of the relative importance of these factors. It is convenient to determine the Reynolds number from the expression $R = LV\rho/\mu$ when L is the length of the system (cm), V its steady velocity (cm s^{-1}) in a medium of density, ρ and viscosity, μ. For fish, R is in the range 10^3 or 10^4, but for microscopic animals it is very small indeed and the stresses due to viscosity are many times greater than those of inertia forces. However, many microscopic organisms, such as flagellates, employ undulatory movements as a means of swimming, as also do spermatozoa. The Reynolds number of a spermatozoon is 3×10^{-6} and inertia of the fluid has no effect on the forces operating against the body.

Undulatory swimming is carried out by animals of diverse grades of tissue organisation, their only common features being an elongate shape. The locomotory waves are retrograde, passing from head to tail in all except the errant polychaetes. Movement of the body is in the lateral plane in fish and polychaetes, the dorso-ventral plane in nemertines and leeches. Species that habitually swim generally have a flattened body so as to increase the area of the propulsive surface. Those animals which are occasional swimmers have to meet the demands of conflicting techniques which give rise to compromises in body form or to the ability to change shape. One example are the leeches, which have a nearly circular cross-section when crawling but become dorso-ventrally flattened when swimming. Another example is the nemertine *Cerebratulus*.[26]

Undulations along edges of the body

In some animals undulatory movements are confined to the lateral margins of the body. This is seen in *Sepia*[18] where movements of the fins can move the cuttlefish gently forward or backward according to the direction of the undulatory waves. The relatively broad polyclad *Leptoplana* is able to swim quite rapidly by muscular ripples which pass along the sides of the worm. The muscular activity concerned is only little different from that used in this and other Turbellaria in muscular creeping (p. 22).

Much more highly specialized developments of this type of swimming may be found in some gastropod molluscs, where the development of the foot have led to the evolution of active pelagic forms. In *Aplysia* and *Pleurobranchus* movements of the side of the foot are comparable to the undulatory waves of ribbon-like animals (Fig. 5.6). These waves can generally be seen to pass along the body

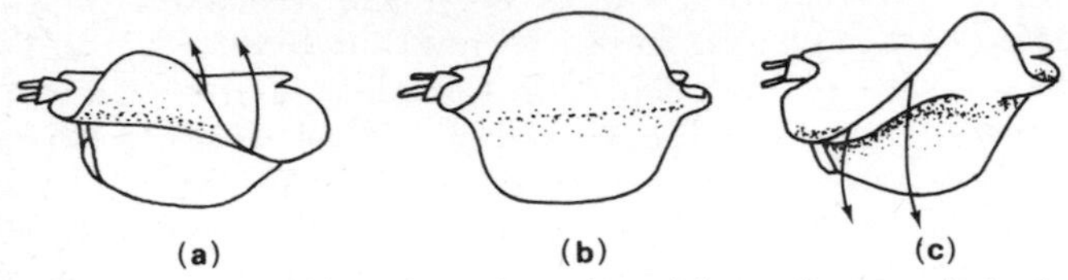

Fig. 5.6 Swimming of *Pleurobranchus* viewed from the morphological right side, gills omitted and swimming lobe of left side shown stationary for simplicity. Arrows indicate movement of lobe. Power stroke on (**c**), recovery in (**a**). (after Thompson and Slinn[107])

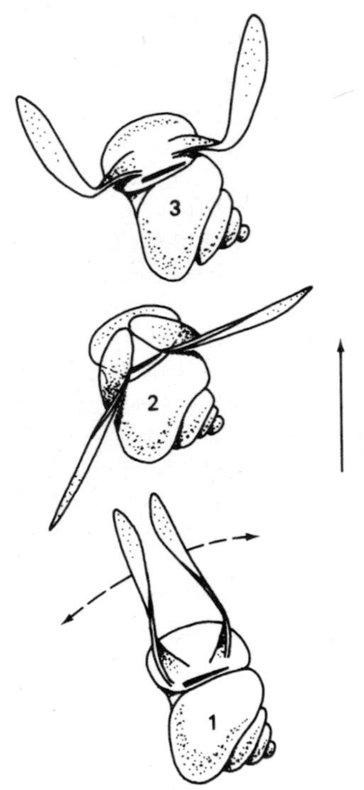

Fig. 5.7 Drawings of *Limacina* showing successive positions of epipodial wings in swimming upwards (arrow). 1 and 2, represent power stroke (broken arrow); 3, recovery stroke. (after Morton[78])

with one or two undulations occurring within the body length (Fig. 5.8) but in *Pleurobranchus* the body is short compared with the wave length of the undulatory movement and propulsion is achieved by distinct, but successive, power and recovery strokes.

The pteropod, *Limacina,* shows even more highly modified structural adaptations and swims by the steady rhythmical beat of lateral epipodia (Fig. 5.7).[78] Here the down thrusting of the wings produces a very effective powerstroke to raise the animal through the water. This is carried out by the contraction of a sheet of muscle fibres passing through the ventral side of the base of each wing. During descent of the pteropod the wings offer no resistance to the water, being held close together above the animal. In the gastropods there has thus been a trend to progress from the passage of a generalized undulatory wave along the sides of the body to the evolution of specialized swimming appendages, such as the wings of *Limacina* or *Clione,* rather than the development of an elongate form specialized for undulatory swimming such as has evolved with metamerism in the errant polychaetes.

The movement of nematodes

Undulatory propulsion occurs in one form or another in all animal phyla and whether the animal is fish, snake or worm the mechanical principles are the same. Such forms as nematodes or spermatozoa, where the body clearly resembles an elongated cylinder, provide the best material upon which a generalized theory can be based.[43]

The body wall of *Ascaris,* in common with that of most nematodes, consists of a cuticle with longitudinal muscles beneath, circular muscles being entirely absent.

Highly vacuolated cells lie between the longitudinal muscles and gut, the vacuoles running together to form what is effectively a fluid-filled perivisceral cavity. Unlike the Turbellaria, Nemertea and the Annelida, there is no antagonism between sets of muscles but between the longitudinal muscles and the cuticle. This contains fibres arranged diagonally in a basketwork pattern and exerts tension against internal pressure. Harris and Crofton[50] have demonstrated internal pressures of as much as 1.6 N cm^{-2}, which accounts for the constant shape and cross-section of nematodes. This pressure, generated by the tone of some longitudinal muscle fibres, may be used to antagonise other fibres, as in undulatory swimming, when dorsal and ventral groups of muscles contract in opposite phase, producing a wave which passes backwards down the body. A consequence of the high internal pressure would be the collapse of the lumen of the gut were it not for a muscular pumping pharynx. The relatively large pseudocoel of *Ascaris* is inconspicuous in small soil dwelling nematodes and the alimentary canal, together with its liquid contents, acts as a fluid skeleton.[101] Movement of gut contents is caused by local contractions of dorsal and ventral longitudinal muscles in phase sequentially along the body provided the cuticle remains at constant diameter. The somatic muscles thicken where locally shortened and radial forces so generated displace the gut.

While an undulating snake traces a winding sinusoidal track over the surface of dry sand so also do small nematodes of 1 mm length (*Haemonchus*) on the surface of agar gel (Fig. 5.8). For the animal to

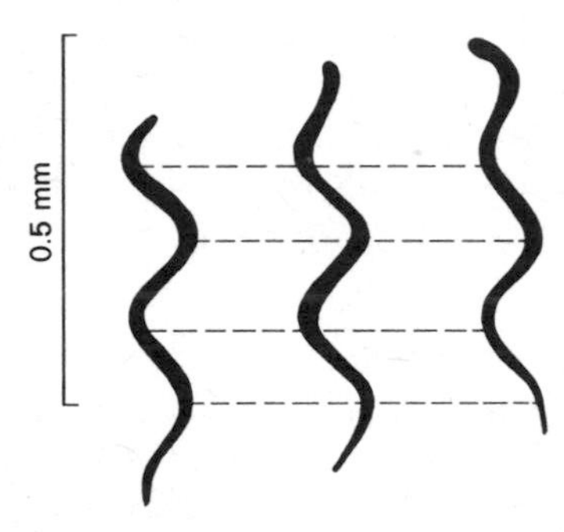

Fig. 5.8 Tracings of *Haemonchus* creeping over agar gel with zero slip, note that the waves are stationary relative to the ground whilst the body moves forward. (after Gray and Lissmann[44])

be able to move forward at the same speed as it is producing waves along its body, each element of the body must move tangentially to its own surface and the substrate must resist any tendency to move under the transverse shearing forces applied by the animal. If the resistance to lateral movement is not sufficient then the track of the animal is wider than the diameter of the body and the animal's velocity (V_x) becomes less than the speed of propagation of the waves (V_w). The waves then travel backwards relative to the ground at a velocity ($V_w - V_x$). The latter represents slip and may be expressed as a percentage of the rate of propagation of the waves. Gray and Lissmann[44] showed that the amount of slip depends on the difference between the coefficients of resistance to movement of elements of the body normal to (C_N) and tangential to (C_L) their own surface. When *Haemonchus* propels itself over the damp rigid surface of an agar gel the slip is 70–80% which indicates that C_N is about twice C_L.

The beet eelworm, *Heterodera*, has been shown by Wallace[124,125] to be able to creep over a damp surface of glass with little or no slip. It is obvious that the air-water interface exerts a profound effect on its movement (Fig. 5.9), there being a critical thickness of water film where the forces resisting slip are maximal. When totally immersed in water *Heterodera* only moves relatively slowly, the viscous forces available to resist the backward passages of locomotory waves only being barely sufficient to produce a reaction to propel the animal forwards. For larger nematodes, such as *Ascaris*, the viscous forces of water are insufficient for forward motion but in a more viscous suspension or in gut contents these animals can move rapidly.

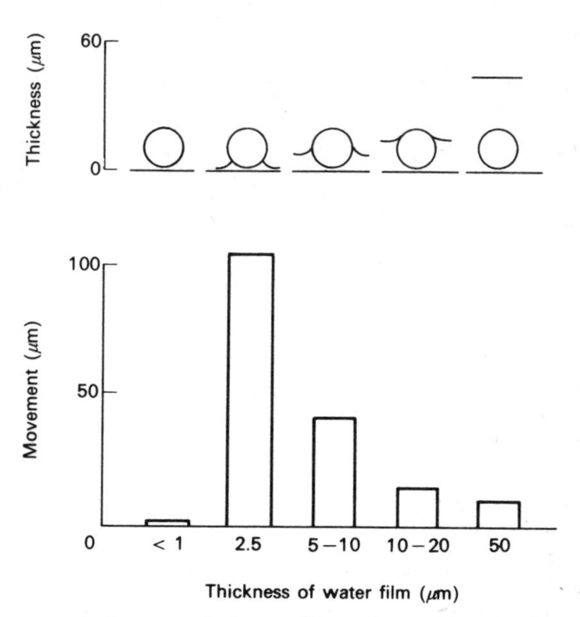

Fig. 5.9 Movement of nematode larvae (*Heterodera*) in water films of different thicknesses. Larvae shown in transverse section in relation to water film above, movement per hour below. (after Wallace[124])

The mechanics of undulatory swimming

This analysis of undulatory swimming may be extended from nematodes to other animals[26,43] When an eel swims, its body is thrown into waves that pass backwards along the body, each short segment moving laterally across the locomotory axis (Fig. 5.10). Segment x travels from right to left and is directed backwards and to the left

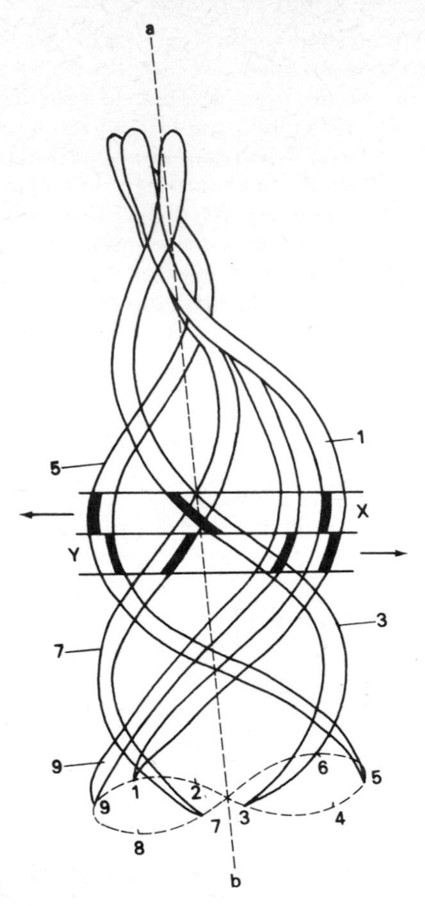

Fig. 5.10 Series of drawings (1–9, some omitted for clarity) of a young eel to show movement of body during passage of a complete wave of undulatory propulsion. a–b indicates locomotory axis; tip of tail describes figure-of-eight curve. (after Gray[43])

during phases 1–5 of the wave; this is followed by movement to the right (5–9) as illustrated by segment *y* which is directed backwards and to the right. Movement of any such segment has two components, one tangential to and the other normal to its surface. The tangential movement exerts no thrust upon the water but the normal component of the motion develops a thrust which results in the animal moving forward. Each section of the body moves at an angle to its surface along the resultant of its tangential and normal

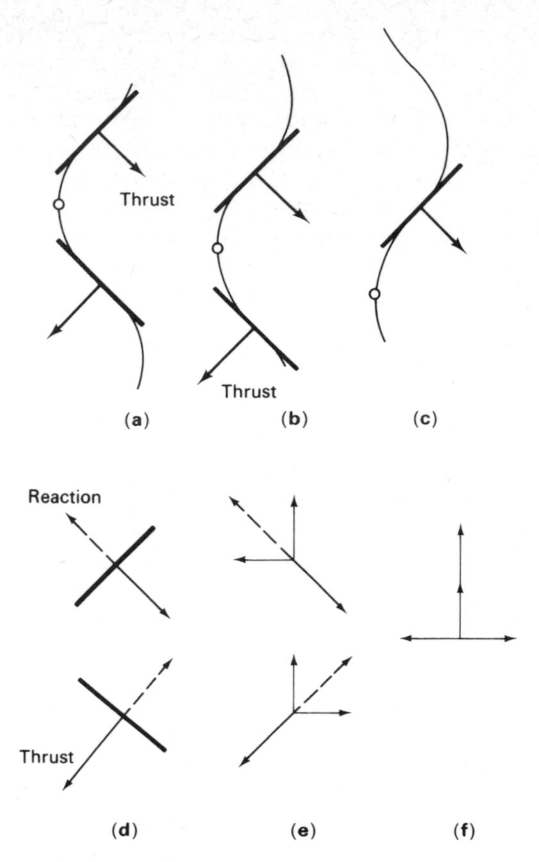

Fig. 5.11 Diagrams to illustrate the forces involved in undulatory propulsion. (**a**), (**b**) and (**c**) show retrograde movement of wave of muscular contraction (circle). Back thrust is exerted by the convex side of each wave and a series of tangents have been drawn to these; (**d**) shows tangents from (**a**) thrust and reaction (broken line); (**e**) resolution of reaction (from **d**) into forward and lateral components; (**f**) vector addition of resolved reactions. (after Chapman[23])

components; thus a smooth animal presents a backwardly travelling inclined surface to the water. A thrust is developed normal to this surface, eliciting a reaction from the medium (Fig. 5.11). This reaction can be resolved into forward and lateral components. The forward components are cumulative and produce forward motion, being equivalent to backward drag on the animal, while the lateral components cancel each other out. A marked characteristic of undulatory propulsion is the increase in wave amplitude as the waves pass posteriorly along the body. This is essential if the resultant transverse forces are to remain zero. If these are not zero the body will tend to yaw from side to side during each cycle of movement. Additionally to avoid yawing, it is essential that more than one complete wave occurs in the length of the body. Relatively short animals, such as *Haemonchus*, accordingly tend to yaw.

The importance of undulatory propulsion is evident from its use in widely different organisms such as snakes, fish, nematodes and leeches as well as in spermatozoa. A significant variation of this locomotory pattern is found in polychaete worms and must now be studied.

UNDULATORY SWIMMING WITH PARAPODIA

Swimming in *Nereis* is essentially an extension of rapid crawling (p.110), the parapodia carrying out backwardly directed power-strokes at the crest of each locomotory wave. In comparison with crawling there is a marked increase in the length and amplitude of the locomotory waves which are direct, passing forwards along the body. This contrasts with other examples described which swim by means of retrograde waves. Indeed, were it not for the paddle-like action of the parapodia *Nereis* would swim backwards. The parapodia turns backwards during their active stroke and so give a forward thrust to the body. However, the undulatory waves pass towards the head faster than the worm moves forwards and their movement might be expected to produce a backthrust on the animal opposing the forward thrust of the parapodia. The solution of this problem was suggested by Taylor[106] who pointed out that analyses of swimming by smooth animals had assumed that the flow of water along the body was laminar but that this ceases to be valid in *Nereis* where the projecting parapodia cause the flow to be turbulent. The above author plotted a series of curves for this condition relating the ratio between forward velocity of the animal and backward velocity of the locomotory wave to the wave form of the undulations. These curves indicated that, unlike the conditions for smooth animals, negative values for the ratio of velocities are possible and correspond to the polychaete condition. Indeed the actual powerstroke of the parapodia are not essential for forward movement; their very presence would generate sufficient skin friction to cause the water flow to be turbulent. This does not mean that the parapodia have no locomotory effect. *Nereis* is a relatively inefficient swimmer whilst *Nephtys*, with a similar arrangement of parapodia, does much better.

The essential difference lies in the manner of functioning of the parapodia. These are perpendicular to the segment surface when the underlying longitudinal muscles are contracted in the trough of the locomotory wave (Fig. 5.12). Indeed during the passage of the trough the recovery stroke is carried out and the parapodium is drawn forward in preparation for the powerstroke. In *Nereis* the powerstroke begins slowly with the parapodium about halfway up the leading edge of the wave and is not completed until the segment is a similar distance down the trailing edge (Fig. 5.12a). *Nephtys* has a powerstroke which is concentrated into a shorter time. It is carried out only at the crest of the locomotory wave (Fig. 5.12b) and

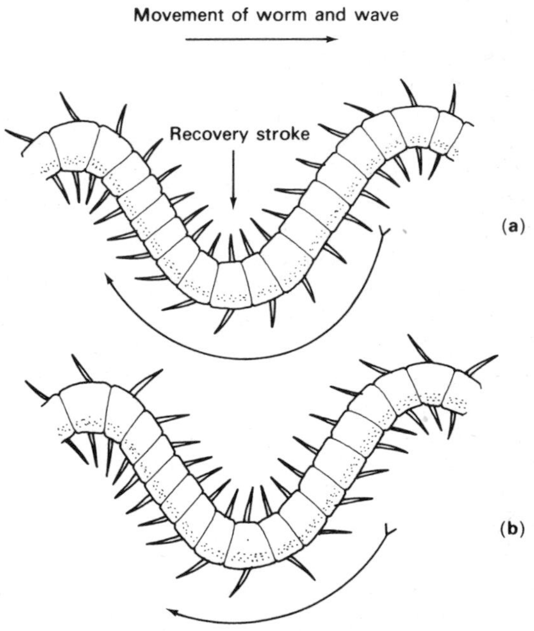

Movement of worm and wave

Recovery stroke

(a)

(b)

Power stroke

Fig. 5.12 Diagrams showing the relation between the inclination of the parapodia and the position of segments of (**a**) *Nereis* and (**b**) *Nephtys*. Duration of power stroke of parapodia indicated by tailed arrow. (after Clark[26])

accordingly the absolute velocity of the parapodium relative to the ground must be much greater than that of *Nereis*.[26] The manner in which the inclination of the parapodium changes in relation to the orientation of a segment during the passage of a locomotory wave

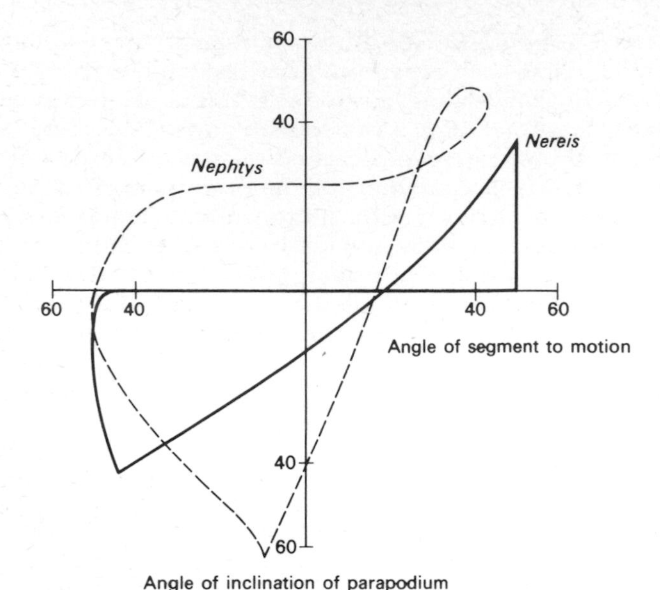

Fig. 5.13 Diagram showing change in angle (degrees) of inclination of the parapodium to the transverse axis of the segment (ordinate) in relation to the inclination of the longitudinal axis of the segment to the direction of motion (abscissa) in *Nereis* and *Nephtys* (broken line) during the passage of a single locomotory wave. (after Clark[26])

during slow swimming can also be shown graphically (Fig. 5.13) and the more rapid powerstroke of *Nephtys* contrasted with that of *Nereis*.

SEGMENTATION AND SWIMMING

Animals of diverse phyla swim by means of undulatory movements and many exhibit metameric segmentation. Questions must now be considered concerning the advantages that a segmental organisation of the musculature may confer, and whether we should look to swimming rather than burrowing for the origins of segmentation. So far as the mechanics of undulatory propulsion are concerned there is no fundamental difference between invertebrates and vertebrates or between segmented and unsegmented animals. All swim in the same manner provided they are of appropriate shape. Differences occur in the means by which the waves are produced according to the basic anatomy of each group. In worms longitudinal muscles are used in a variety of locomotory mechanisms, one of which is swimming. Undulatory movements can be made whether the muscles are segmentally arranged or not. Bending of the body is carried out by contraction of longitudinal muscles along one side and the fluid pressure changes are transmitted to other parts of the body wall, causing elongation of the antagonistic muscles and flexure of the body. The force of muscle contraction is transmitted by the fluid skeleton to the body wall directly, not by means of the septa. Segmentation is not essential, and indeed smooth transmission of locomotory waves down the body is more easily accomplished by a series of overlapping longitudinal muscles rather than by their being restricted by segmental boundaries.

When we consider the chordates, particularly the more primitive ones, for example *Branchiostoma*, the lancelet, longitudinal muscles are used exclusively for the generation of swimming movements. These muscles are arranged segmentally and this appears to be fundamental to the performance of undulatory swimming in association with the development of a notochord or axial skeleton.

In amphioxus and in fish the longitudinal muscles occur in a series of successive blocks or myotomes, each characteristically bent into a zig-zag pattern. The axes of the muscle fibres are all approximately parallel to the axial skeleton and their contraction would induce very little bending if the intermuscular septa or myocommata were perpendicular to the long axis of the fish, for the pull would be almost entirely exerted against the adjoining myotomes. However, the myocommata are inclined to the perpendicular and so permit a transverse component of the force of contraction of the myotomes to exert a torsional force on the axial skeleton. Views on the evolution of segmentation in the chordates are summarized by Clark[26] who reaches the conclusion that chordate metamerism evolved as an adaptation to undulatory swimming.

A good example of this means of locomotion in the chordates is seen in amphioxus (*Branchiostoma*) in which the waves of lateral displacement of the body increase in amplitude towards the tail during forward swimming. Amphioxus has a notochord consisting of a collagenous sheath enclosing a system of transverse paramyosin muscular plates, the sheath and myocommata forming a continuous structure (Fig. 5.14). Guthrie and Banks[45] have demonstrated that the notochord functions as a hydrostatic skeleton, and that tension may be developed in the myocommata, under electrical stimulation, so as to affect the stiffness of the notochord. This stiffness is under

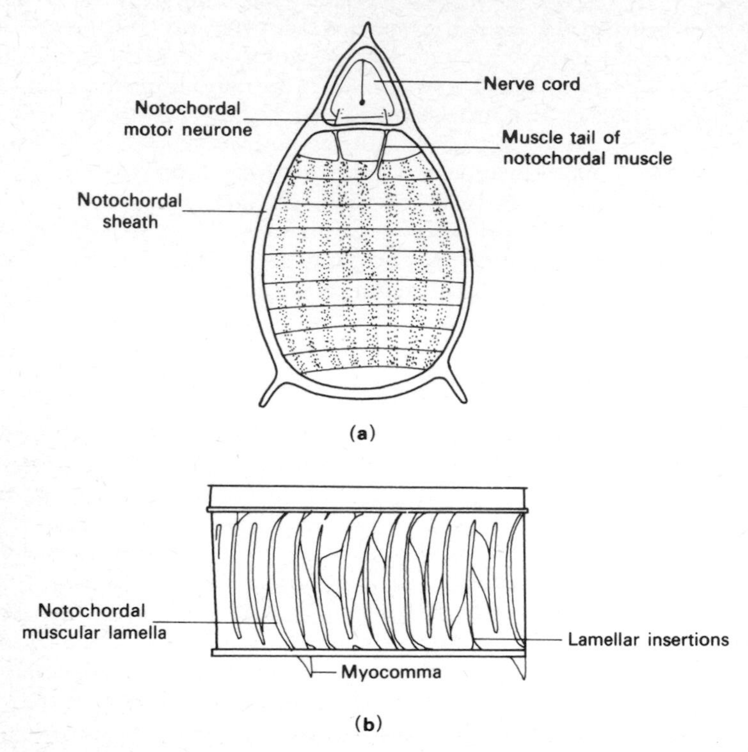

Fig. 5.14 The structure of the notochord of *Branchiostoma*. (**a**) transverse; (**b**) longitudinal section. (from Guthrie and Banks[45])

control of the nervous system for the notochordal muscles contract with the myotomal muscles, the connections between spinal cord and notochord being described by Flood.[36] The varying stiffness of the notochord is particularly important in respect of fast undulatory swimming in which the myotomes may contract more than 20 times each second, applying torsional force to the notochord by means of the myocommata, so as to bend the body through angles of 10–44°. In slow swimming the range was 22–110°, suggesting a more flexible notochord at slow speeds. Guthrie and Banks used a preparation of the notochord to study passive and active resistance to flexure (Fig. 5.15). The active tension component contributes most to the total resistance to flexion at low angles; 63% at 9° deflexion, 50% at 16°, falling to 25% at 45°. This active increase in stiffness at the low angles of flexion, characteristic of fast swimming, is especially important in

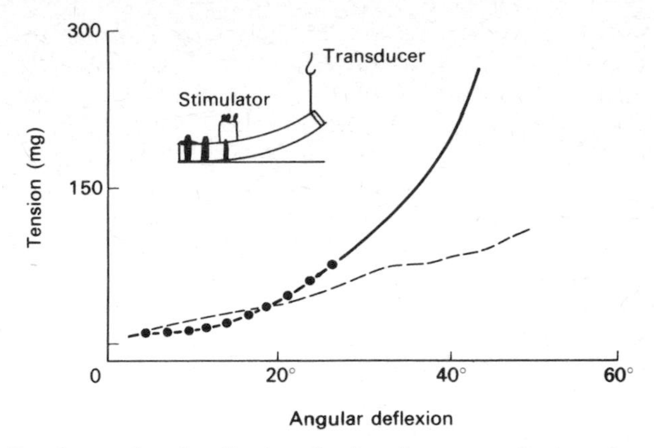

Fig. 5.15 Passive and active (broken line) resistance to flexion of notochord preparation (inset). Notochord held down on left, tension recorded by transducer (T) and active response obtained by direct application of electrical stimulation (s) at 10 Hz. (after Guthrie and Banks[45])

providing elastic recoil of the notochord. In slow swimming, with flexions of up to 110°, the contribution of the active component will be very small.

Amphioxus also possesses the rare ability to swim tailwards by means of lateral displacement of the body. Webb[128] describes how it may come to rest, when swimming forwards, and reverse direction extremely rapidly. Undulatory waves are now generated in the tail, instead of in the head as in forward movement. The undulatory waves increase in amplitude as they pass along the body, reaching a maximum displacement at tail or head in forward or reverse locomotion respectively. The control of this amplitude is probably a function of the notochord, for contraction of its muscular plates would affect the stiffness and so the amplitude of undulatory movement.

The ability to reverse would be of value in the avoidance of predators while swimming, but Webb considers that its chief importance is probably in movement through sand. Head or tail first entry, according to the end touching the substrate, ensures rapid burrowing, whilst the ability to emerge from sand by reversal is comparable to the behaviour of some burrowing bivalves, for example *Donax* (p. 85).

The manner in which the force of contraction of the lateral muscles of fish or *Branchiostoma* is applied to the axial skeleton by myocommata as a torsional force is entirely different from the manner of functioning of the longitudinal muscles in annelids. Tension produced by the contraction of longitudinal muscles in worms is never transmitted by the septa but by means of the fluid skeleton, which thus has an entirely different role from the fluid skeleton of the notochord. It is now generally accepted that segmentation of the chordate body evolved independently of the annelids as a means of undulatory swimming with an axial notochord. By contrast in an annelid metamerism is not essential for swimming to take place and we must consider further the reasons for segmentation in this group when we come to discuss the principles of metazoan evolution.

This example of undulatory swimming in amphioxus is of particular interest in respect of the notochord. Here is what appears to be a classical chordate axial structure and yet it has now been shown to function as a fluid-muscle system. It superficially resembles the arrangement in Nematoda; both have a tough enveloping membrane and a single set of muscles, transverse in the notochord and longitudinal in nematodes. Both have evolved as fluid-muscle systems suited for undulatory swimming independently in quite separate phyla. This suggests that the similar mechanical problem of skeletal flexure in undulatory swimming has resulted in similar solutions. We shall consider how far this obtains in respect of other problems of locomotion in respect of metazoan evolution in Chapter 7.

6

Swimming by Jet Propulsion

INTRODUCTION

We have so far been concerned with locomotion over or into substrates, and with undulatory swimming, but it is by the use of jet propulsion that soft-bodied animals achieve their most striking successes. Indeed, it is in association with jet swimming that the cephalopod molluscs attain the zenith of invertebrate development both convergent to and comparable with the evolution of the bony fish.[87] This means of swimming is characteristically a development of soft-bodied animals, for the method is only exploited elsewhere in the animal kingdom by dragonfly larvae[76], and by fish which develop a forward thrust when water is forcibly ejected from the gill chambers. Many fish at rest have to compensate for this thrust by continued movement of the pectoral fins, whilst predators, such as the pike, utilize jet propulsion by the rapid expulsion of water from the gills for increased acceleration from rest.[61]

The essential requisite for jet swimming is a body form which is sufficiently deformable to allow a considerable amount of water to be forcibly expelled and which possesses adequate elasticity to expand after contraction, taking in fluid so that a further power stroke may be delivered. Thus in any animal swimming by jet propulsion, one would expect the fluid-muscle system to be especially well developed, and the Mollusca with their double system of mantle cavity and haemocoel to be at an advantage. In the opisthobranch, *Notarchus*,[73] and in the scallops[77] it occurs largely as an escape mechanism. In other groups, for example in jellyfish, squid and salps, jet swimming has evolved further into a highly effective means of locomotion.

In all animals utilizing jet propulsion, a common basic principle is applied: the expulsion of water in one direction to propel the animal in the opposite direction. Clearly an animal would be driven furthest if as large a volume of water as possible is expelled, as in an escape reaction, when a sudden movement away from a predator may be crucial. In normal swimming, however, this would be wasteful of energy. Alexander[2] indicates how swimming may be most economically accomplished by the movement of a large mass of water at low velocity rather than by a small amount of water at high velocity. This is substantiated in the squid (*Loligo vulgaris*), where the pressure generated in the mantle cavity during normal swimming is about one tenth of the maximum pulse.

We must turn to the cephalopods to examine the principles in more detail for in this group jet propulsion is developed *par excellence* and it is here that it has been investigated in most detail. Two aspects of the problem merit particular attention: firstly, how the mantle cavity of these molluscs is adapted to produce jets of water, the pressures developed, and the dynamics of their locomotion; secondly, the mechanics of refilling of the mantle so as to enable a series of jet cycles to occur in rapid succession.

JET PROPULSION IN THE CEPHALOPODA

Anatomy of the mantle cavity

In the squids, cuttlefish and octopus the mantle cavity lies ventrally and to a limited extent laterally to the viscera in the posterior region of the body (Fig. 6.1). The mantle has developed into a thick wall of muscular tissue of which a large part consists of circular muscles whose fibres are orientated in the plane of the section shown in Figure 60b. It is the contraction of these fibres that causes water to be expelled from the mantle cavity during respiration or jetting. In addition there are radial and longitudinal muscle fibres.

There are separate inhalant and exhalant apertures at the interior edge of the mantle, the inhalant being paired and lateral to the central single exhalant funnel-like opening (Fig. 6.1c). The funnel may be pointed in various directions (Fig. 6.6) and is attached to the body by a pair of powerful retractor muscles. Together with the head retractors, these muscles are essential to hold the body together during jetting. Flow of water in the reverse direction is

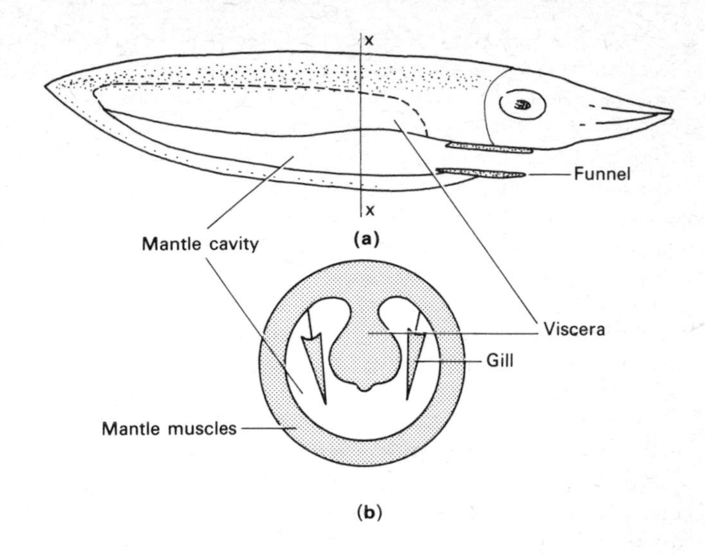

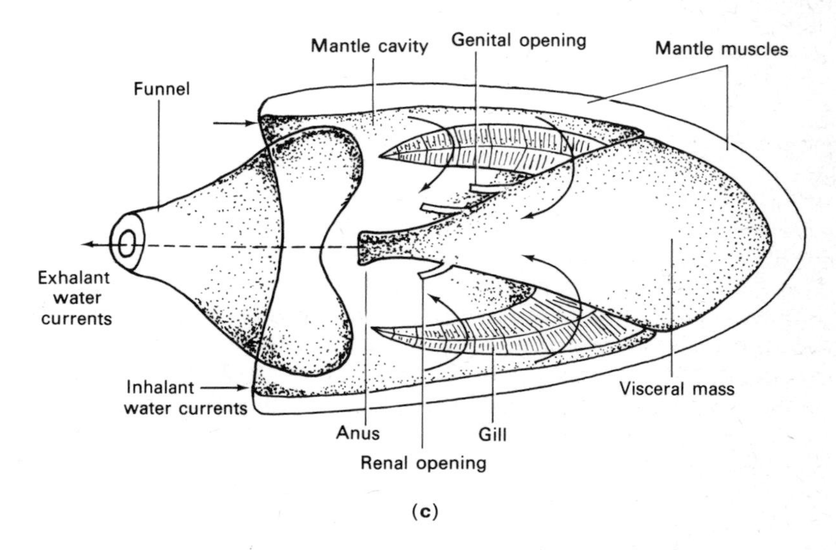

Fig. 6.1 Diagrams of *Loligo* (**a** and **b**) and *Sepia* (**c**) to show the mantle cavity, viscera and mantle muscles. a, Sagittal section, behind the head; b, transverse section along line x–x in a; c, ventral aspect of mantle cavity with the central part of the mantle muscles removed.

prevented by valvular mechanisms. These consist of a simple flap of tissue in the funnel, preventing inhalation of water, and the outer collar of the funnel, which is distended by water pressure in the mantle cavity and locks into the cartilagenous sockets of the mantle, one in each inhalant aperature. These valves can be readily observed to control the direction of water flow in *Sepia* and *Loligo*. Water is drawn into the mantle cavity laterally, passes through the gills and over the rectal and excretory openings, and is exhaled through the central funnel.

The upper surface of the mantle cavity bears accessory reproductive glands and consists of only a thin layer of little more than the dermal and coelomic epithelium to separate the mantle cavity from the extensive renal and pericardial coelomic cavities containing the heart and viscera. Thus any high pressures generated in the mantle cavity by jetting probably also affect the coelomic fluid. Innervation of the mantle muscles is by means of the paired stellate ganglia from which third order giant neurones cause a simultaneous massive contraction of the circular muscle fibres of the mantle to develop a high pressure pulse in the mantle cavity characteristic of jetting. Fine nerve fibres, giving a graded response, are thought to cause low amplitude pressure pulses including respiratory mantle movements.[88]

Analysis of jet propulsion

It is apparent from the application of Newton's laws of motion to jet swimming that the expulsion of a mass of water (m) at a given velocity (q) will impart momentum to a squid (mq) in the opposite direction. The resultant velocity of the squid depends upon its residual mass, that is its mass after the water has been expelled from the mantle cavity, and upon its drag. Maximal momentum may be imparted either by an increase in jet velocity or by an increase in the amount of water expelled. The former involves the generation of greater mantle pressures by the development of more powerful muscles and the latter an increase in the capacity of the mantle cavity. Both of these factors may be seen to be highly developed in *Loligo* (Table 6.1).

In a hydrodynamic analysis of jet swimming consideration must be given to the forces associated with gravity, drag, acceleration and jet thrust. The inertial forces concerned with acceleration depend on the total mass of the body, together with that of the water remaining in the mantle cavity at any instant, and upon the instantaneous acceleration ($\mathrm{d}u/\mathrm{d}t$) of the body. The thrust de-

Table 6.1 Comparison of mantle capacity and musculature of some cephalopods in order of increasing swimming ability (after Trueman and Packard[123]).

	Mantle capacity as % of body weight	Mantle muscle weight as % of body weight	Average mantle thickness (cm)
Octopus vulgaris (370 g)	10	<10	0.22
Sepia officinalis (250 g)	20–30	30	0.75
Loligo vulgaris (350 g)	>50	35	0.52

veloped by the jet is given by the product of jet velocity (q) and the rate of flow (dm/dt) through the funnel. Johnson, Soden and Trueman[55] have deduced a simple expression for calculating the increase in velocity (ΔU) resulting from a single jet pulse:

$$\Delta U = q \ln(1 + m_c/m_b) \tag{1}$$

where m_b is the mass of the body and m_c the total mass of water expelled. However, it has been assumed in deriving this formula that gravity and drag forces are negligible and that the jet velocity is constant throughout its discharge. *Loligo* is only about 4% more dense than sea water so that it is reasonable to consider gravity as having little affect but the drag factor is by no means negligible nor does the jet velocity remain constant during discharge.

The rate of mass flow through the funnel can be regarded as the product of jet velocity, the outlet area of the funnel (a), and the fluid density. Losses occur at the funnel during discharge and may be allowed for by the introduction of a coefficient of discharge (C_D) which is one or less, although we have little direct evidence as to its value. However, on this basis the previous authors deduced an expression for jet thrust:

$$q(dm/dt) = 2C_D \, . \, a \, . \, P \tag{2}$$

Thus both jet velocity and jet thrust can be calculated if the instantaneous pressure in the mantle cavity can be determined. Since jet thrust was determined experimentally with the animal held stationary, by being tethered, drag need not be considered and a direct comparison between jet thrust and pressures may be made.

Experimental observations

Difficulties were encountered in recording the pressure generated by squid because of the rapidity of jetting and the problems of handling a free swimming squid in a large tank. This problem was largely solved by tethering the animal with a nylon thread stitched through the cranial cartilages, after light narcosis, or by a small metal hook attached to the base of the arms. A pressure cannula was then carefully inserted in the mantle cavity through an inhalant opening (Fig. 6.2).[123] With this technique it was possible to obtain

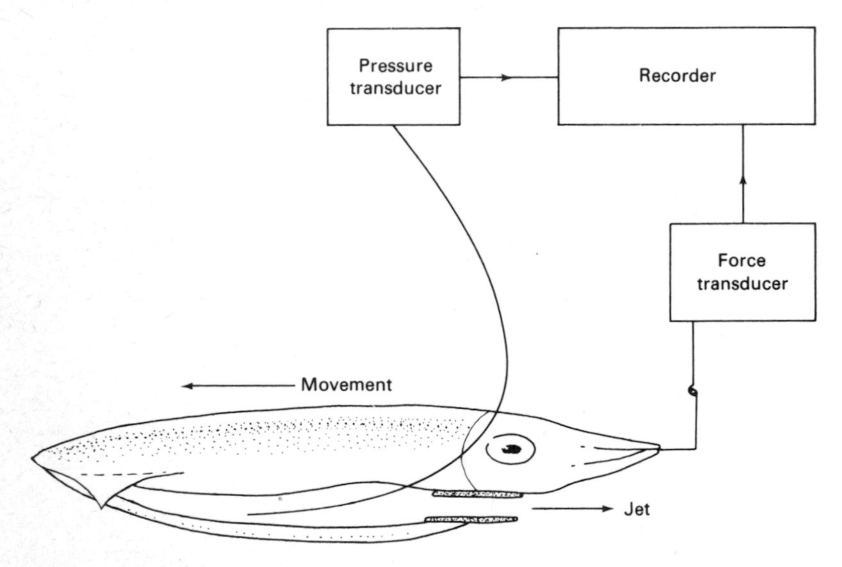

Fig. 6.2 Representation of the technique used to restrain squid and record jet pressures and thrust simultaneously. The animal is tethered by a thread attached between a hook inserted near the mouth and a force transducer.

recordings of the pressure changes in the mantle cavity of a variety of cephalopods both at rest and during jet swimming (Fig. 6.3a). At rest a regular low pressure respiratory pulse is apparent. This rhythm is broken by jet pulses which are of reduced duration with increasing amplitude (Fig. 6.3a). High pressure pulses were recorded from the mantle cavity of all cephalopods investigated, a maximum of 4 N cm^{-2} being obtained from *Eledone* of 600 g wet weight. All maximal pressure pulses are notable for their sharp rise and fall, having a duration of about 180 m sec in *Loligo*, for example. A period elapses after all high pressure pulses during which refilling

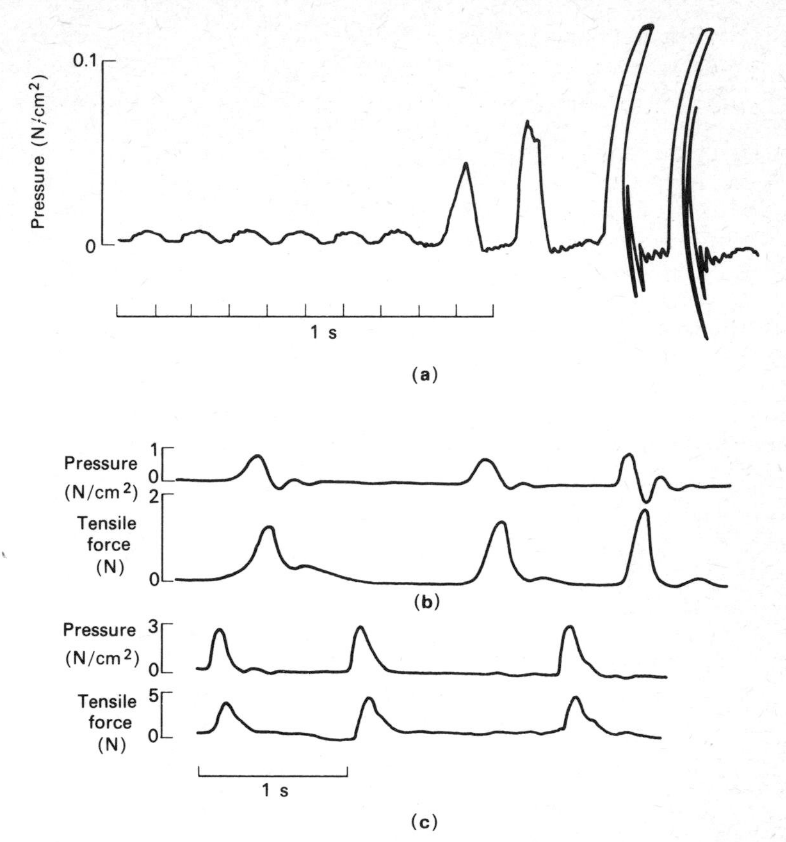

Fig. 6.3 Pressure recordings (N cm^{-2}) from within the mantle cavity of *Sepia* (**a** and **b**) and *Loligo* (**c**). In **a**, the respiratory rhythm is broken by the development of jets of increasing amplitude and decreasing duration. Oscillations occurring after the final two pulses are due to the recording system used; **b** and **c**, are simultaneous recordings of pressure and the tensile force (N) developed in the thread as shown in Figure 6.2. (From Trueman and Packard[123])

of the mantle cavity takes place. This pattern of a power stroke followed by recovery occurs cyclically in respect of each jet pulse and is referred to as a jet cycle. No distinct negative pressures were observed during the refilling of the mantle but those may have been obscured by the oscillations occurring after pressure pulses for technical reasons on most traces.

Simultaneous recordings of pressure and of jet thrust were readily made by attaching the thread, used for tethering the animal, to a force transducer (Figs. 6.2 and 6.3b and c). The maximum thrust developed in *Sepia, Loligo* or *Eledone* is approximately equivalent to their body weight, although it should be pointed out that the thrust recorded may be in part due to swimming using the fins. This probably accounts for the irregularities in the base line for thrust in *Sepia.* The quantitative relationship between pressure and tension developed can be more clearly seen in graphs derived from the data of numerous recordings similar to those figured here (Fig. 6.4). In each the broken line represents the condition of jet thrust = $2C_D \cdot a \cdot P$ (equation *2*) where C_D is one, and all species perform less well than this. However, the alignment of the animal with the axis of the force transducer is critical for maximal values of thrust and some unusually low values may be accounted for in this

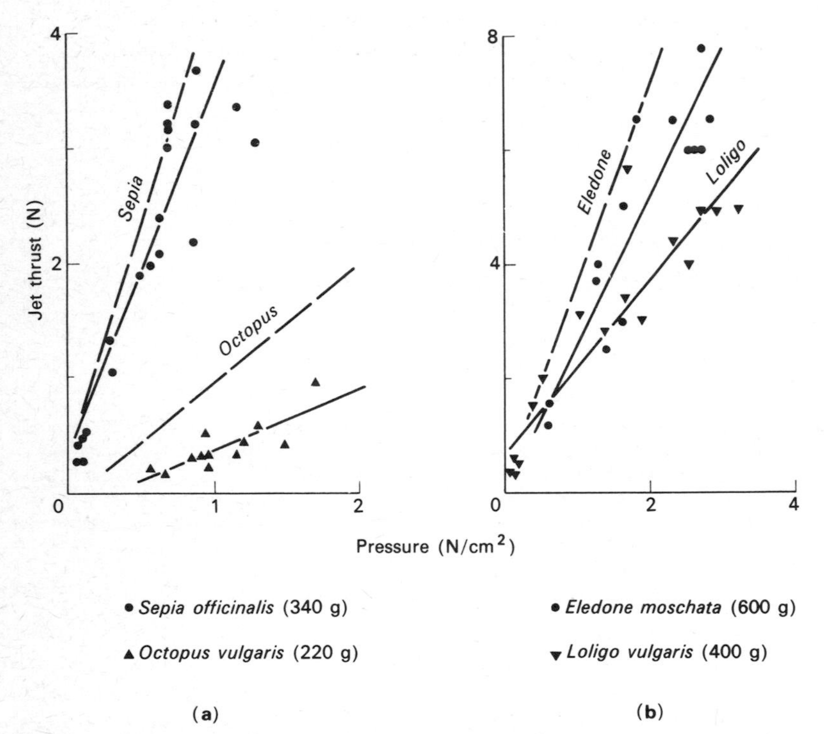

Fig. 6.4 Graphs showing relation of mantle cavity pressure and jet thrust (determined as tensile force) in **(a)**, *Sepia officinalis* and *Octopus vulgaris*; **(b)**, *Eledone moschata* and *Loligo vulgaris*. Regression lines determined by method of least squares, broken lines indicate relationship jet thrust = $2C_D \cdot a \cdot P$ where $C_D = 1$ for the animals named (after Trueman and Packard[123]).

way. The area of cross-section of the jet (a) is derived from measurements of the diameter of the funnel in the living or freshly dead animal, but the aperture alters during the development of the jet in a way in which it has not been possible to assess. It is thus not practicable to use these observations to determine a more realistic value for the coefficient of discharge of the funnel (C_D). The relationship, expressed by equation 2, between thrust and pressure is closest in *Sepia* and least apparent in *Octopus*. The latter is probably caused, in part, by the constriction of the funnel during jetting and in part by the octopus reacting to being tethered in a different manner to other cephalopods. *Octopus* produces, for example, a pressure pulse of longer duration but of lower amplitude, and undoubtedly has a poor swimming performance when compared with *Loligo*.

The jet thrust produced by a series of small *O. vulgaris* of different body weight was compared with their ability to hold onto the side of the aquarium with five arms (Fig. 6.5). An octopus of about 1 g body

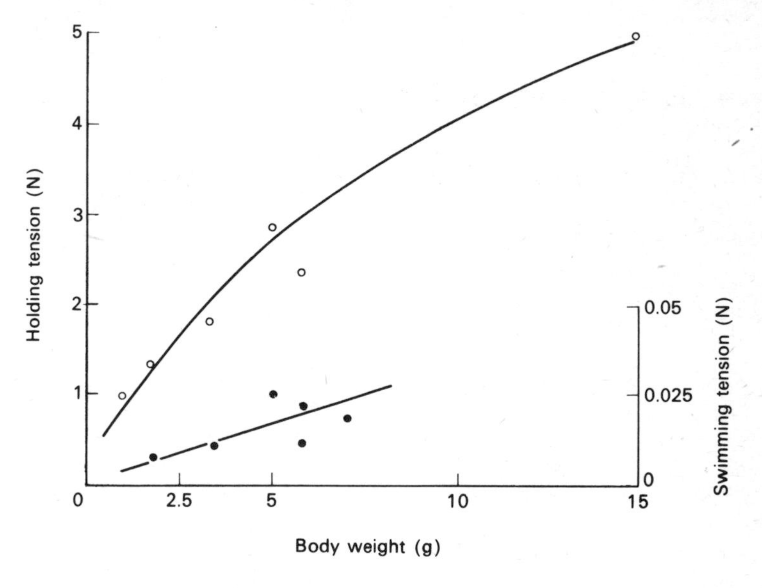

Fig. 6.5 Graphs showing relationship between the increase in the body weight of young *Octopus*, the maximum tensile forces developed by production of jets and the maximum holding pull with five arms attached to the tank. The latter represent points at which a thread stitched through the cephalic cartilage broke free (from Trueman and Packard[123]).

weight produces a jet thrust of less than 1×10^{-2} N but a pull of 1 N. This latter represents a muscle tension of 20 N cm^{-2} when the pull is considered in relation to the cross-sectional area of the longitudinal muscles at the base of the arms. This development of the muscles of the arms rather than of the mantle reflects the essentially bottom-living habit of the octopus (Fig. 6.6). A comparison as regards mantle capacity and musculature in adult cephalopods is summarized in Table 6.1, in which the animals are arranged in order of swimming ability.

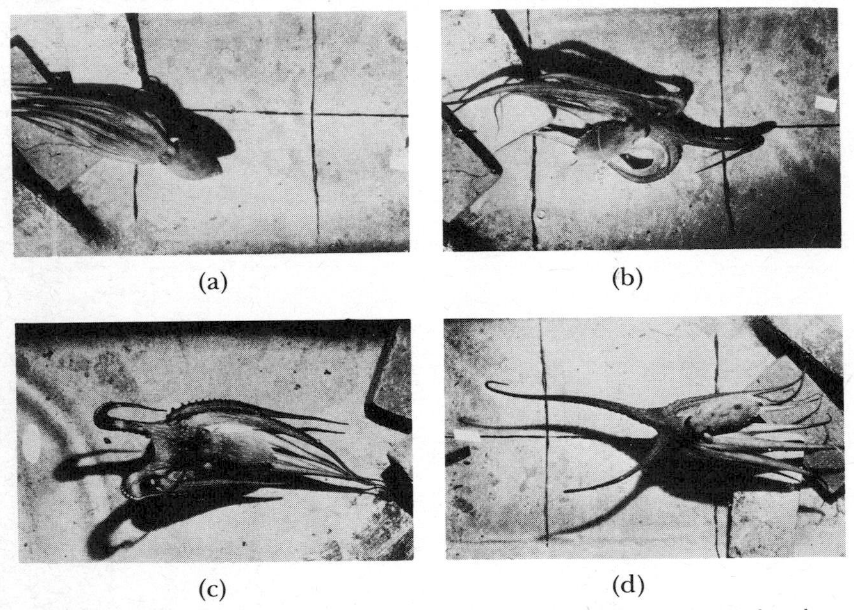

(a) (b)

(c) (d)

Fig. 6.6 Photographs of *Octopus vulgaris* attacking white discs. (**a**) jet swimming backwards, note stream lined shape; (**b**) gliding over the substrate using powerful arms; (**c**) and (**d**) jet swimming forwards with two arms extended in attack, remaining arms trailing, note funnel reflected posteriorly beneath left side of head (photographs, A. Packard).

It may be recalled that the attainment of high swimming velocity by jet propulsion is closely related to the velocity and mass of water ejected from the mantle cavity. These are affected by the following: (a) the thickness of the mantle muscle and its contractile properties, for these determine the pressure that may be generated; (b) the cross-sectional area of the funnel, for jet velocity is inversely proportional to this; and (c) the volume of water contained within the mantle cavity. The first two factors both affect jet velocity, the latter the mass of water ejected. In normal locomotion it is most efficient to expel as large a volume as possible with low mantle cavity pressures and in consequence low jet velocity. The high pressure pulses recorded in *Loligo* are normally associated with rapid escape movements and the squid is commonly exhausted after making a number of these. In contrast, normal swimming of *Loligo* consists largely of rather deep respiratory pulses of the order of 10×10^{-2} N cm^{-2}. These are economical of effort and may continue over long periods. But, however a cephalopod may swim, mantle

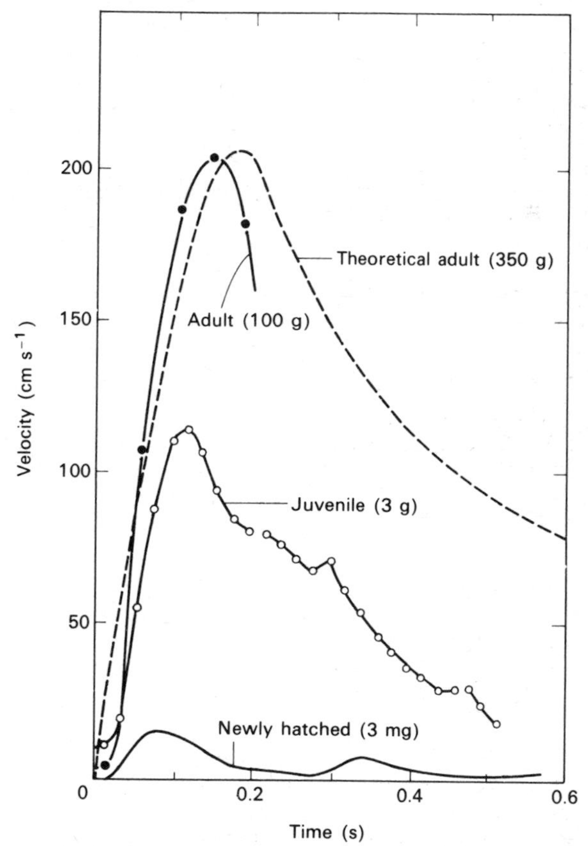

Fig. 6.7 Theoretical and experimental curves showing increase in velocity for *Loligo vulgaris* of different weight during a single jet cycle. Experimental data derived from high speed photography (Packard[86]), theoretical curve derived from mathematical model (from Johnson, Soden and Trueman[55]).

cavity capacity is critical. If the capacity is small then the pressure pulse must be of short duration or low amplitude. There is clearly adaptive value in having a large mantle capacity and powerful muscles such as the relatively thick mantle muscles occurring in *Loligo* and *Eledone*. Jet propulsion favours animals with large body size and it has been suggested that this accounts for the rapid rate of growth of the cephalopods.[85]

Stroboscopic photography has been used by Packard[86] to determine the velocity of *Loligo* during a single jet cycle from a standing start (Fig. 6.7). Young squid with body weights of 3 mg, 3 g, and 100 g achieve progressively greater maximal velocities in < 50, 50–60 and 85–125 m s respectively. These times are comparable to the time for maximal tension to develop in a mantle nerve-muscle preparation of *Loligo* and represent, in a 100 g squid, an acceleration of 3·3 *g*.

Comparison of squid performance with a simple theoretical model

The analysis of motion of *Loligo vulgaris* has been developed further with the aid of a mathematical model, the equations used for this purpose being detailed by Johnson, Soden and Trueman.[55] A squid is considered as a hollow sphere of uniform wall thickness containing water in which the proportions and related properties in a 350 g adult *Loligo* were used as far as possible. For example, the volume of water taken to be expelled from the sphere is the same as ejected from the mantle cavity of *Loligo* (200 ml) and the funnel outlet was taken to be the same area as observed by Trueman and Packard.[123] No information was available on losses in the funnel on discharge so that the discharge coefficient (C_D) has been assumed to lie in the range 0·6–1. These data and other assumptions implicit in the model enabled graphs to be drawn showing the theoretical relationship between time and mantle cavity pressure, jet thrust, jet velocity and distance moved during a single jet cycle. These all showed reasonably good agreement with the observations of pressure and jet thrust. Theoretical velocities for squid were also estimated for single jet cycles with allowance being made for drag. It was assumed that the drag coefficient remained constant at an arbitary value of 0·47 regardless of squid velocity (this is the value for a sphere of the same size and velocity as a squid). Squid density was assumed to be the same as seawater so as to eliminate gravity effects and hence the need to consider direction of motion. On the basis of this model; and with a coefficient of discharge of 0·6, a maximum speed of about 200 cm s^{-1} is deduced for a single jet cycle. This velocity is much less than that previously deduced, 330 cm s^{-1}, by use of equation (*1*) without consideration of drag, and it corresponds with the speed observed by Packard.[86] A theoretical curve showing acceleration of a 350 g squid is superimposed upon Packard's results (Fig. 6.7). Similar calculations have been made using a model of cylindrical shape rather than spherical. These indicate that the ability to generate jet thrust is not altered by shape, but that an elongated form may considerably reduce the drag component.

The effect of drag during jet swimming will be to reduce the maximum velocity and to cause it to occur earlier in relation to the pressure pulse. The reason for this may be understood by consideration of the forces acting on the body (Fig. 6.8). In the absence of drag and gravity forces the squid would continue to accelerate until the jet thrust falls to zero, but if drag forces are present the squid will cease to accelerate as soon as drag forces equal jet thrust. The jet thrust falls as soon as mantle cavity pressure decreases and drag then causes retardation even before the pressure pulse is complete. Consideration of the effect of different losses in the funnel have

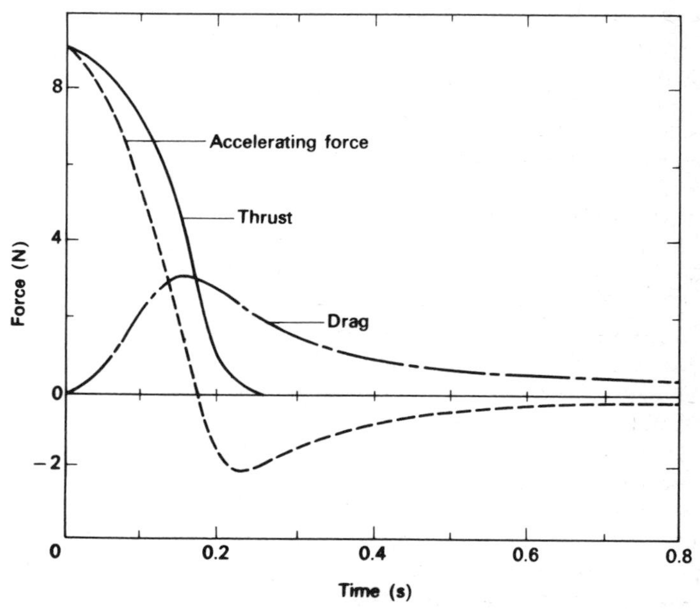

Fig. 6.8 Graph summarizing the forces acting on the body of a squid as a result of a single jet cycle (from Johnson, Soden and Trueman[55]).

shown that when at their lowest ($C_D = 1$) the instantaneous velocity is highest but the squid does not travel so far in 0·8 s. This is due to the combined effects of a shorter pressure pulse duration and higher velocity leading to greater drag losses.

The use of this model has recently been extended from consideration of single jet cycles to multiple jet cycles. A series of cycles have been assumed to occur at appropriate intervals of 1·5 s. Each cycle gives a large positive pressure pulse on jetting followed by a small negative pulse during the refilling of the mantle. The effect of refilling, and of drag developed as a consequence of high speed, is to reduce the velocity between cycles. After the first cycle, however, jets occur whilst the squid is in motion and a higher velocity is consequently attained. Maximal velocity is achieved after only 4 or 5 cycles with a value of 350 cm s^{-1} for *Loligo vulgaris* of 350 g.

It is very likely that this sort of multijet situation obtains when squid are avoiding predators, Lane[62] records an occasion when *Loligo pealeii* was pursued by fast fish and momentarily jumped out of the water. There are also other reports of the hooked squid, *Onchoteuthis banksii,* landing on the decks of ships and of having been observed to fly through the air.

MANTLE MUSCLES DURING THE JET CYCLE

Mantle anatomy

One of the outstanding problems of jet propulsion is the refilling of the mantle cavity but before we can discuss the mechanism by means of which this may be accomplished we must first consider the anatomy of the mantle.

This description is based on the accounts of Ward and Wainwright[126,127] and upon personal observations. The squid mantle consists essentially of outer and inner tunics of collagen fibres arranged in a trellis pattern at an angle of 27° to the long axis of the mantle (Fig. 6.9a). These are thought to function in a similar manner to the collagen fibres in the body wall of nemertine worms[28] but whereas the volume contained in a worm remains constant, the mantle cavity volume changes considerably. The role of these collagen fibres is probably to maintain the length of the mantle constant, and to prevent any disintegration of the mantle which might follow the sudden and massive contraction of the circular muscles when generating a high pressure pulse in the mantle cavity. The space between the collagen tunics is full of deformable, yet incompressible, muscle tissue which functions in contraction of the

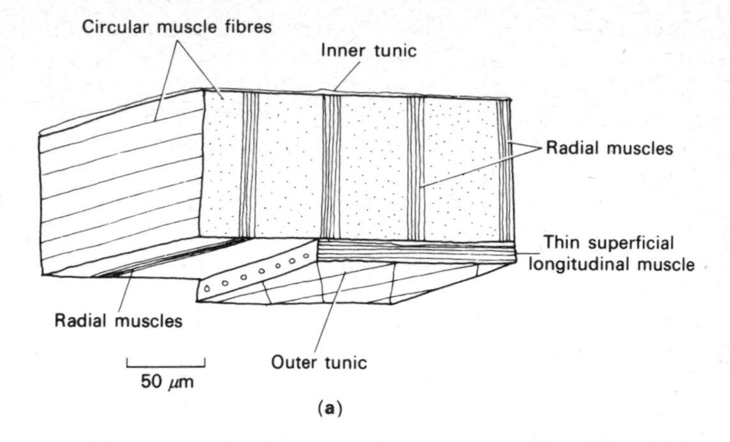

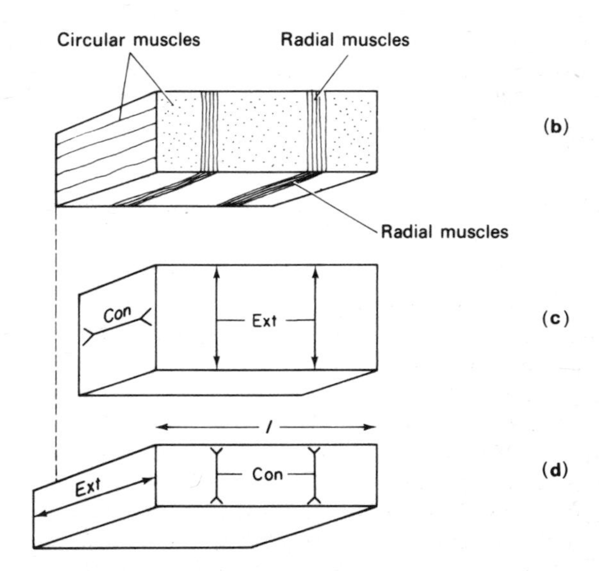

Fig. 6.9 Diagrams of the structure of the mantle musculature. (**a**) shows a small block of mantle of *Sepia* with outer and inner collagenous tunics, a thin superficial layer of longitudinal muscle, circular muscle fibres with radial muscles passing in bands transversely through the mantle; (**b**), (**c**) and (**d**) show a similar block of the mantle in different phases of contraction. (**b**) resting stage; (**c**) circular muscles contracted (con >—<) to produce a jet pulse, radial muscles extended (ext ←→); (**d**) radial muscles contracted, circular muscles extended during hyperinflation of the mantle. b–d are drawn at constant length (*l*) and each contain the same volume. The thickening and thinning of the mantle is thus accompanied by the contraction and extension of the circular muscles as indicated by the broken line.

mantle and also as the resistant fluid of the hydroskeleton of the mantle. Cylinders are supportive systems whose resistance to buckling depends on the rigidity of the material, the radius of the cylinder and the thickness of its walls.[2] The rigidity of the mantle wall is due in part to the collagen tunics, in part to tension in the mantle muscles.

The circular muscle is the principal component of the mantle. Passing through the mantle in the radial plane, lie radial muscles whose fibres form sheets separating the circular muscle into blocks. In *Sepia* (Fig. 6.9a) a thin layer of longitudinal muscle lies outside the circular muscle, being particularly evident near the mantle margin, but these longitudinal fibres are apparently absent from squid.[127] All these muscles must be supplied with blood from the mantle arteries, but no haemocoelic spaces have been observed within the mantle tissues. Thus no great volume of extra-muscular fluid can participate in the mechanism of antagonism between the muscles that causes expansion and contraction of the mantle. An exception occurs in the cranchid squids where large blood-filled cavities are located in the mantle. These cavities play an important role in achieving neutral buoyancy.

Changes in the mantle during jet cycles

Observation of the dimensional changes in the mantle of squid or cuttlefish during respiration or jet cycles proved to be a difficult technical problem for it was necessary to monitor the animals under as nearly normal conditions as was possible. Successful recordings were made by inserting electrodes in the mantle, under light anaesthesia using urethane, and by passing a high frequency current (25 KHz, 2 μ A) between them.[88] Movement of the electrodes apart or together increased or decreased the inter-electrode impedance respectively and afforded a means of recording movements of the mantle. The animal was meanwhile confined in a small net. The electrodes used were either a pair of needles inserted parallel to each other and orientated so as to indicate shortening and extension of the circular muscles, or a pair of button electrodes located one on either side of the mantle to monitor changes in thickness (Fig. 6.10b).

Simultaneous use of these electrodes, located near the anterior margin of the mantle of *Sepia*, showed that during respiration, exhalation involved the shortening of the circular muscles and the thickening of the mantle, and that during inhalation, the circular muscles lengthened and the mantle became thinner. These changes

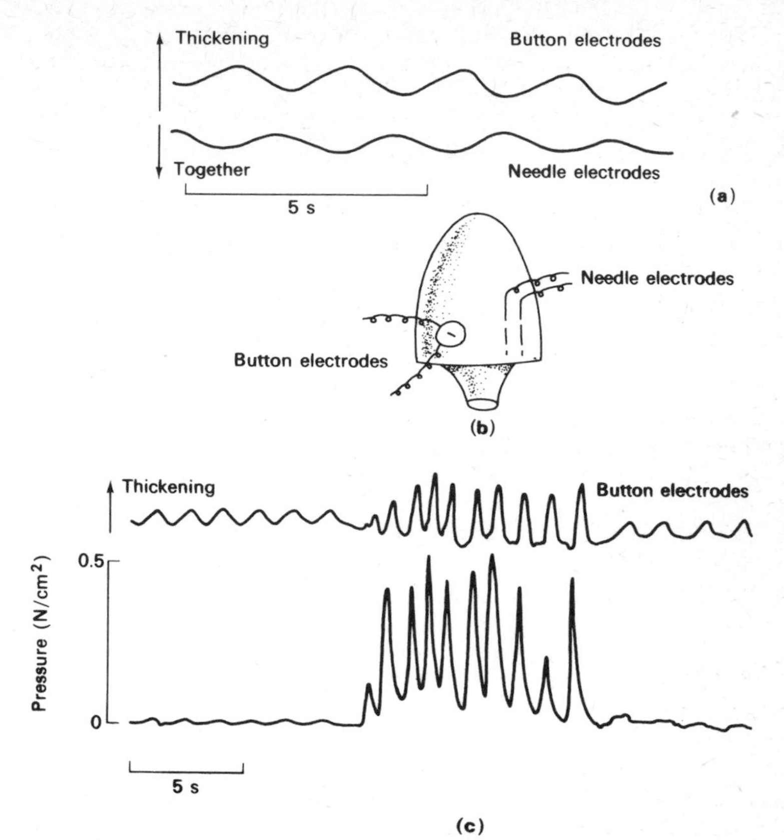

Fig. 6.10 Recordings of mantle contractions during respiration (**a**, in *Sepia*) and jet cycles (**c**, in *Loligo*). In **a**, button and needle electrodes are used simultaneously to show thickening of mantle, contraction of circular muscles bringing the needle electrodes together, the electrodes being placed as in **b**; **c**, pressure pulses recorded with the thickening of the mantle during respiration followed by series of jet cycles (after Packard and Trueman[88]).

were even more pronounced during jet cycles when, in *Loligo*, a decrease of about 10% of the circumference of the mantle was observed (Fig. 6.10). Use of needle electrodes in the posterior and anterior regions of the mantle of *Sepia officinalis* simultaneously indicated that, in this species at least, respiration is confined to the anterior region but that the posterior mantle is certainly involved in the production of high pressure pulses (Fig. 6.11). During respiration the mantle appears to contract from what may be termed the resting condition, but prior to the generation of a high pressure

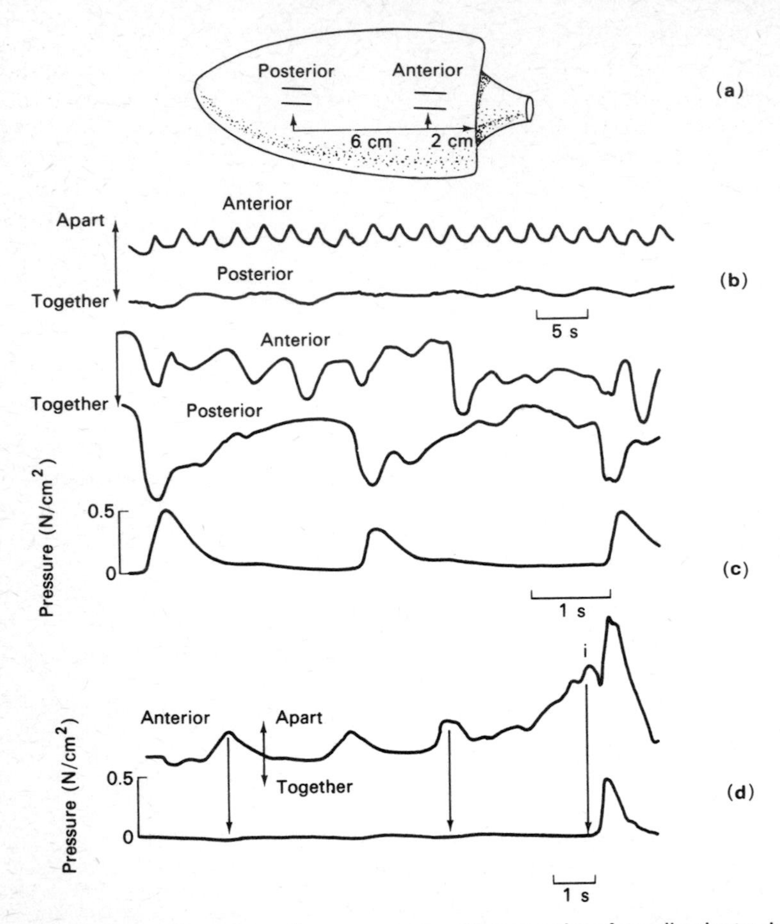

Fig. 6.11 Recording of mantle movements with two pairs of needle electrodes inserted superficially in the mantle muscle as shown in (**a**); (**b**) Respiratory contractions of circular muscles only in anterior electrodes; (**c**) major contraction of anterior and posterior of mantle with high pressure pulse; (**d**) contraction and relaxation of anterior mantle in respiratory rhythm followed by hyperinflation (i) causing the electrodes to move much further apart immediately prior to high pressure pulse (after Packard and Trueman[88]).

pulse the animal could on occasion be observed to hyperventilate the mantle cavity. This involved the thinning of the mantle from the resting condition and its hyperinflation (Fig. 6.11d). This was further observed in *Sepia*, for the mantle was expanded during jet cycles, even when in air, to the extent of appearing like a small balloon. This was further investigated by means of a mantle nerve-muscle preparation in which a flap of muscle tissue was cut from the mantle leaving one margin attached so as to retain innervation from the stellate ganglion. The free edge of the flap was attached to a movement transducer which enabled changes in the length of the circular muscles to be observed during jet cycles (Fig. 6.12). These recordings clearly show three conditions of the circular muscle corresponding to the resting condition and to contraction and extension beyond the resting condition.

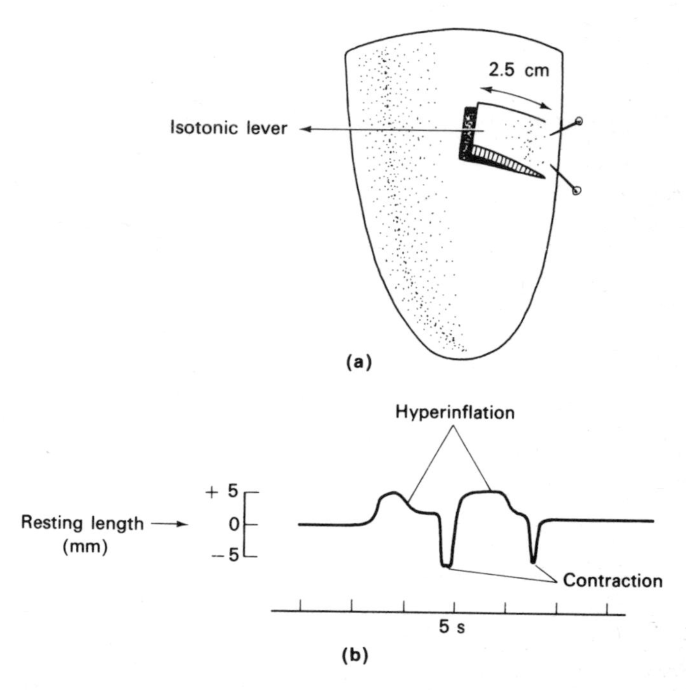

Fig. 6.12 Mantle flap preparation (**a**) of *Sepia* in which changes in circular muscle dimension can be monitored directly by isotonic lever while the flap is held by pins and retains innervation. (**b**) typical recording showing hyperinflation from resting length during jet cycle followed by contraction.

Particular attention was also paid to the length of the mantle, but no change could be discerned during respiration and only very small changes during jet cycles. This confirms Ward's[126] observations made when analyzing cine film of *Loligo*. It is probably correct to assume that the antero-posterior length of the mantle remains constant during the contraction and extension of the circular muscles. This may in part be caused by the collagen tunics and in part be brought about by the superficial longitudinal muscles (Fig. 6.9a). If the changes in dimension of a cube of mantle tissue are now considered at a constant overall mantle length (Fig. 6.9b, c, d), contraction of the circular muscles results in their thickening so as to occupy the same volume, this being the only dimension in which they are able to expand to compensate for shortening. As the circular muscles thicken so they must stretch the adjacent radial muscles. Recovery of the circular muscles is brought about by the contraction of the radial muscles, causing thinning of the mantle, provided again that the overall mantle length remains constant. Indeed, the further contraction of the radial muscles is apparently the force causing the elongation of the circular fibres beyond their resting condition during hyperinflation of the mantle. Figure 6.11 demonstrates that the extension of the circular fibres is by active muscular action.[88] It is clear that radial and circular muscles act in mutual antagonism in a similar manner to the circular and longitudinal muscles of a worm. But whereas in a worm forces are brought to bear between muscles by means of the coelomic fluid, no comparable fluid-filled cavity is found in most cephalopod mantles. Muscular antagonism in the cephalopod mantle involves direct interaction between each band of circular muscle with its adjacent sheets of radial muscle. The fluid skeleton is represented by the muscles themselves, rather than by any fluid-filled cavity. Such a system is basically a localized one in which forces are not transferred over great distances, but the intimate association of all bands of circular muscles with two sheets of radial fibres allows active expansion of the whole mantle. This arrangement of muscle fibres facilitates the rapid inhalation of water so essential for a rapid succession of jet cycles.

It may be added in passing that this arrangement of mantle muscles is simple compared with that of the cephalopod arms and tentacles. Here again, in the absence of fluid-filled chambers, muscle must antagonize muscle directly to allow the arms to be used with fine co-ordination.

Control of the mantle muscles

The distribution of giant neurones from the stellate ganglia to the mantle muscles has been well established.[132] The largest diameter fibres conduct impulses most rapidly and pass to the most remote (posterior) part of the mantle, whereas thinner, more slowly conducting fibres innervate the muscles of the anterior region of the mantle. This arrangement allows all the circular muscles of the mantle to contract simultaneously to produce a pressure pulse and this was confirmed by monitoring thickness changes of the mantle during jet cycles. Any arrangement of neurones whereby the posterior circular muscle fibres contracted after those situated more anteriorly is quite impracticable, for constriction of the mantle in the region of the funnel would impair the outflow of water and might lead to the rupturing of the mantle wall.

The leading edges of pressure pulses are at their steepest in recordings of maximal pressure pulses. In *Loligo* and *Sepia* they are almost certainly brought about by a giant fibre response, characteristic of an all or nothing contraction system, the time taken for the mantle to contract being constant and independent of the amount of shortening of the muscle fibres. When low pressure pulses are developing the pressure rises more slowly, as shown by the first and third pulses in Figure 6.3a. Low pressure pulses are probably caused by fine nerve fibres giving a graded response and this suggests that the circular muscles may be doubly innervated. Another feature of low pressure pulses is their longer duration, pulse widths being progressively shorter with rising amplitude. The volume of water in the mantle cavity is limited and with higher pressures greater jet velocities result in the mantle cavity being emptied more rapidly.

The innervation of the mantle muscles is complex, not only because of double innervation but also because of the distribution and function of different mantle muscles. Circular and radial muscles require to be reciprocally stimulated and inhibited so as to act antagonistically, while longitudinal muscles must be under tension whenever the former muscle fibres contract in order to prevent lengthening of the mantle. Little is known of the pattern of distribution of neurones to circular and radial muscles or about the co-ordination of their contraction. Investigation of muscle action potentials in the mantle by Ward[126] supplied little evidence to resolve this problem owing to the difficulty of being certain from which group of muscles the potentials were arising. Two possible systems of muscular control would be: (a) separate innervation for each group of the circular, radial, longitudinal muscles, (b) common innervation for all mantle muscles. While the first suggestion is perfectly reasonable, it would require a relatively complex system of neurones. A simpler mechanism involving the simultaneous stimulation of these three groups of muscles could, however, function, provided that the response time and the decay of tension in the different muscle groups was appropriate. Immediate contraction and rapid decay of tension in the circular muscles, with a longer lasting response in the longitudinals, coupled with a slower but longer lasting response in the radial muscles, might provide a functional basis. Hyperinflation of the mantle with excess contraction of only the radial muscles, however, indicates that the innervation of the mantle is more complex than the latter alternative suggests.

JET PROPULSION IN OTHER SOFT-BODIED INVERTEBRATES

Notarchus

Although the cephalopods are jet swimmers *par excellence*, it should not be overlooked that this means of locomotion also occurs in the other two major molluscan groups. *Notarchus* is a sea hare (Opisthobranchia) in which the two lateral parapodia form a nearly closed sac, or parapodial cavity, which on contraction ejects a propulsive water jet (Fig. 6.13). The result is jet swimming in a peculiar looping manner[73] (Fig. 6.13b and c). Water is taken into and expelled through the same parapodial opening with sufficient force to lift the animal off the bottom over potential predators. The force of this jet is applied in a downward direction, imparting motion in a wide curve. When the jet stops a fast somersault begins, consisting of a single backwards rotation until the parapodial opening faces downwards. This is apparently the stable equilibrium position with the viscera concentrated at the anterior (lower) end of the body adjacent to the parapodial opening. During rotation the parapodial aperture opens and the cavity becomes refilled with water, so that the next power stroke begins with the aperture again pointing downwards. The somersault undoubtedly occurs because the sea hare has no fins with which to steer its movements; thus the water jet propels it into a state of unstable equilibrium and from this it rapidly turns downwards when the jet ends.

In the laboratory some animals have been recorded as continually swimming for as long as 7 min, jet cycles occurring at the rate of 3 in 7 s with each jet lifting them approximately 10 cm through the water. However, little detailed information is available concerning the pressure developed, the jet thrust, or the mechanism of refilling of the fluid-filled sac. This swimming activity seems to be simply an escape reaction from crabs and sea anemones rather than the more

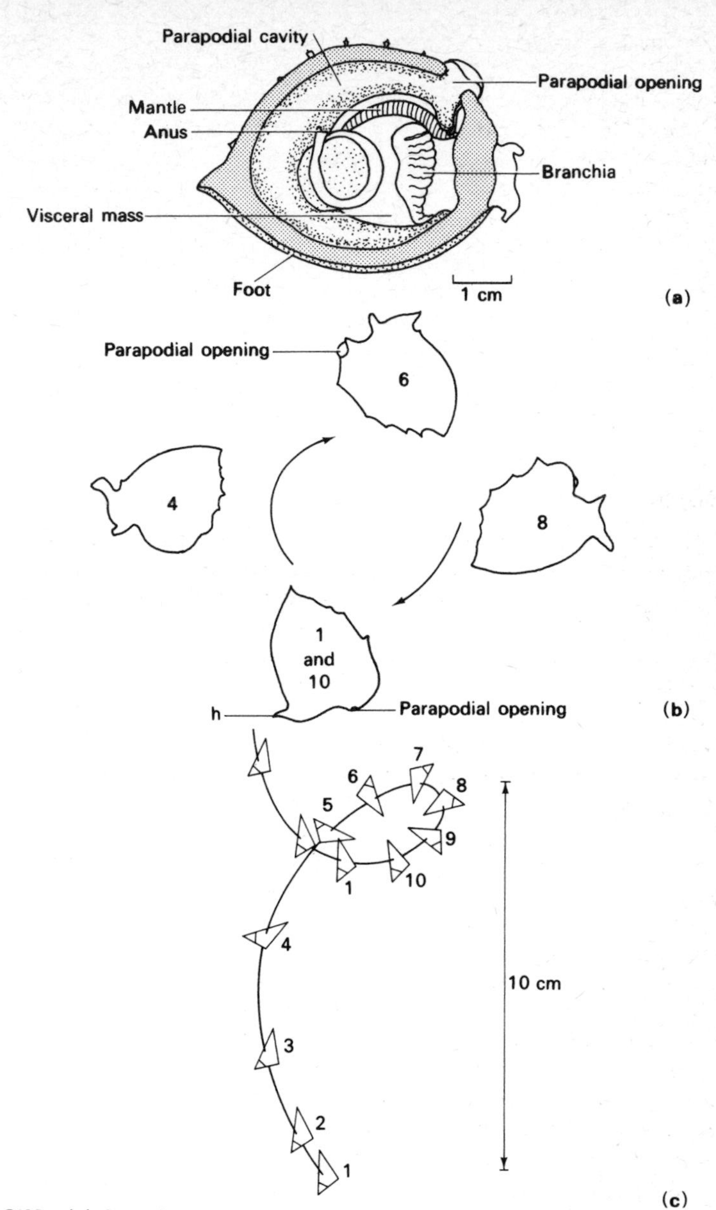

Fig. 6.13 (**a**) A sagittal section through *Notarchus punctatus*. (**b**) Positions assumed by *Notarchus* during swimming, numbers corresponding to those in **c**, in 1 the parapodial opening forms a narrow canal through which water is expelled; in 6, the opening is wide for the refilling of the parapodial cavity. (**c**) Reconstruction of swimming from cine film; time interval 0.4 s. Propulsive (1–5) and rolling (6–10) phases shown (after Martin[73]).

normal method of locomotion of crawling by pedal locomotory waves.

Scallops

Water jets from the mantle cavity of bivalves play an important role in burrowing but it is only in the scallops (Pectinacea) that such jets have been used extensively in swimming. Jet swimming is carried out by comparatively small modifications of the mantle, shell, and adductor muscles characteristic of all bivalve molluscs. These changes have been comprehensively reviewed by Yonge.[133] Swimming is achieved by the repeated flapping of the valves; the

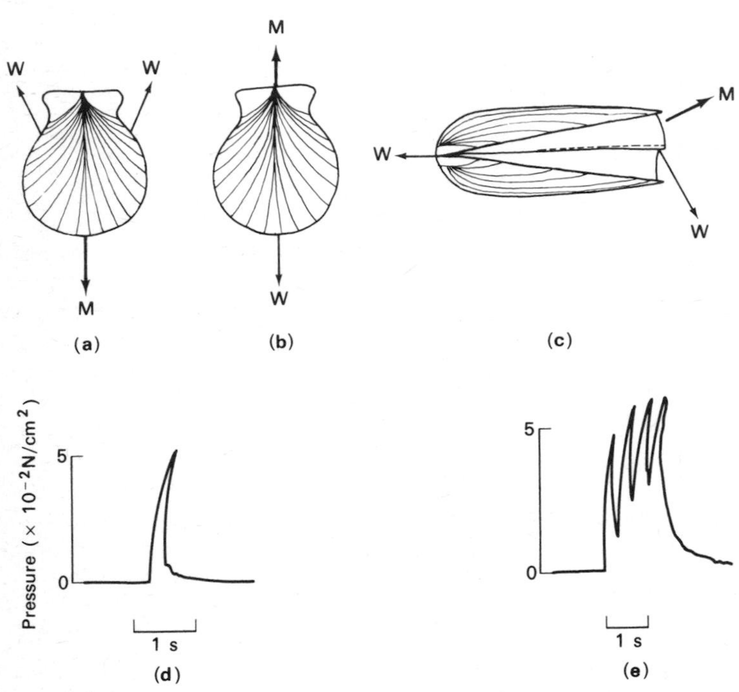

Fig. 6.14 Swimming movements of the scallop showing expulsion of water jets (w, small arrow) and resulting movement of the animal (M, large arrow). (**a**) normal swimming, movement being in the ventral direction; (**b**) escape swimming, hinge foremost; (**c**) normal swimming from the side showing expulsion of water between velar folds of the mantle to give 'lift'; (**d**) pressure pulse derived from a single adduction of the valves of *Chlamys opercularis* during swimming; (**e**) series of pulses caused by repeated adductions during normal swimming (after Moore and Trueman[77]).

water ejected raises the scallop off the bottom and often propels it on a somewhat erratic course. In normal swimming, scallops move with the ventral margin of the valves foremost because of the production of paired jets in the opposite, i.e. dorsal direction (Fig. 6.14a). The mantle flap or velum of the upper (left) mantle lobe passes over that of the lower, so that some water is forced downwards to cause an upward thrust (Fig. 6.14c). Tactile stimulation of the mantle, or the presence of a starfish near the ventral margin of the valves, results in what has been termed an escape movement. The velum is drawn back so as to allow a single jet to pass out from between the ventral mantle margins. This results in the scallop swimming with the hinge foremost, often on a very erratic course characteristic of a random escape movement (Fig. 6.14b).

The pressures developed in the mantle cavity, while producing the water jets, have been recorded by means of a pressure transducer coupled to the cavity by plastic tubing. In *Chlamys opercularis* pressure pulses of 5–6 and 3×10^{-2} N cm^{-2} were recorded during normal swimming and escape movements respectively (Fig. 6.14d). Immediately prior to adduction the valves would open about 4° more widely than normally, to a maximal gape of about 20°. This draws more water into the mantle cavity for use in the jet and is directly comparable to the hyperinflation of the squid mantle. Following adduction, water is drawn into the mantle cavity as the valves reopen. This occurs when the adductor muscles relax because of the potential energy stored in the rubber-like hinge ligament.[1,109] Single jet cycles are commonly produced less frequently than a series of cycles. These occur in rapid succession with the jets directed either in the dorsal or ventral directions. The mantle pressure does not fall to zero between adductions, particularly during normal swimming, but rather builds up in a step-like manner (Fig. 6.14e). During normal swimming, movement is in a ventral direction and, as the valves open, water is forced into the mantle cavity. Thus the velocity gained from each jet cycle serves, in part, to refill the mantle cavity, the inflow of water allowing the build-up of pressure.

Jet velocity (q) may be calculated either from the volume of water expelled (v) in known time (t) in a jet of cross-section area (A), $q = v/At$, or from mantle cavity pressure (P) records using the Bernoulli equation, from which the maximal velocity of the jet is $\sqrt{2P/\rho}$ where ρ is the density of the water. *Chlamys* expels 26 ml of water at each adduction, determined by measuring surface area of valve and angular change in gape at adduction, and with a pressure pulse of 6×10^{-2} N cm^{-2} a jet velocity of about 115 cm s^{-1} is obtained using either method.[77] For escape movements a pressure pulse of

3×10^{-2} N cm^{-2} would generate a jet velocity of 57 cm s^{-1}. The expression (equation *1*, p. 133),

$$\Delta U = q \ln(1 + m_c/m_b)$$

for calculating the increase of velocity of a squid during a single jet cycle, has been applied to the scallop. The theoretical maximal velocities for a single jet cycle are 46 and 16 cm/s for normal and escape swimming respectively. Comparable determinations from cine film gave velocities of 25 and 13 cm s^{-1}. The differences are probably due in part to the formula not taking into account the effect of gravity or drag forces. It has also been assumed that all the water from the mantle cavity passes out through the jet orifices, but it is recognized that during normal swimming some water is forced downwards between the mantle margins to cause an upward movement. If one third of the mantle water is taken to pass out in this manner then the theoretical maximal velocity is reduced from 46 to 30 cm/sec, a value comparable with that determined from film of normal swimming. It should be additionally noted that the scallop is relatively poorly streamlined and often has incrusting growths on the shell so that, as in the squid, drag may cause deceleration even before the pressure pulse has terminated. Another point is that normal swimming is more rapid than the escape movement, but the ventrally directed jet may have a dual role by thrusting away the would be predator as well as causing evasive movement.

Yonge [133] considered that the adaptations of the Pectinacea, enabling them to swim, are no more than developments of those originally acquired for the efficient cleansing of the mantle cavity. The periodic rapid closure of the valves of bivalves is thus a preadaptation to swimming. The rubber-like hinge ligament contributes to the ability of members of this group to sustain swimming movements by the repeated flapping of the valves with minimal energy loss, for these ligaments are outstanding as the most efficient mechanically in the Bivalvia. The angle through which scallops close their valves is particularly wide in relation to other bivalves[116] and allows more water to be expelled. Compared with the pressure pulses recorded from bivalves that burrow (p. 68), the pulses from scallops seem at first sight to be surprisingly low in amplitude. But further consideration suggests that scallops practice jet propulsion efficiently by expulsion of a maximal volume of water at relatively low pressures and in consequence low jet velocities. This is more economical of energy than accelerating a smaller volume of water to a higher velocity.

Jellyfish

Jellyfish, medusae and the swimming bells of siphonophores are another good illustration of jet propulsion. Once again the same basic principle of displacement of water occurs, in jellyfish by means of subumbrella muscles. The most detailed study has been made by Gladfelter[39] on the hydrozoan, *Polyorchis*. This animal is cup-shaped and swims upwards by means of a downwardly directed jet expelled through the velum (Fig. 6.15). The jet is brought about by a

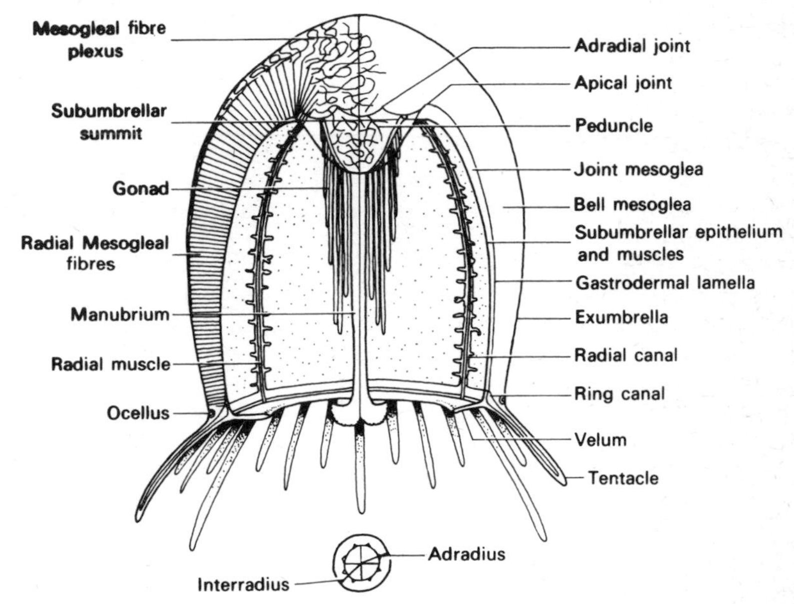

Fig. 6.15 Longitudinal section through the bell of *Polyorchis montereyensis*; interradial section (left) and adradial section (right) as in small diagram below. (from Gladfelter[39])

reduction in the diameter of the sides of the bell by contraction of the subumbrella muscles, while the upper region of the bell remains relatively constant in diameter (Fig. 6.16). The recovery phase of the jet cycle draws water into the subumbrella cavity as the original shape is restored by the elasticity of the mesoglea. The mesoglea stores part of the energy of contraction of the subumbrella muscles which is released as these muscles relax, the mesoglea thus having the same role as the hinge ligament in scallops. This mechanism, which is described in detail by Gladfelter,[39] involves contraction of

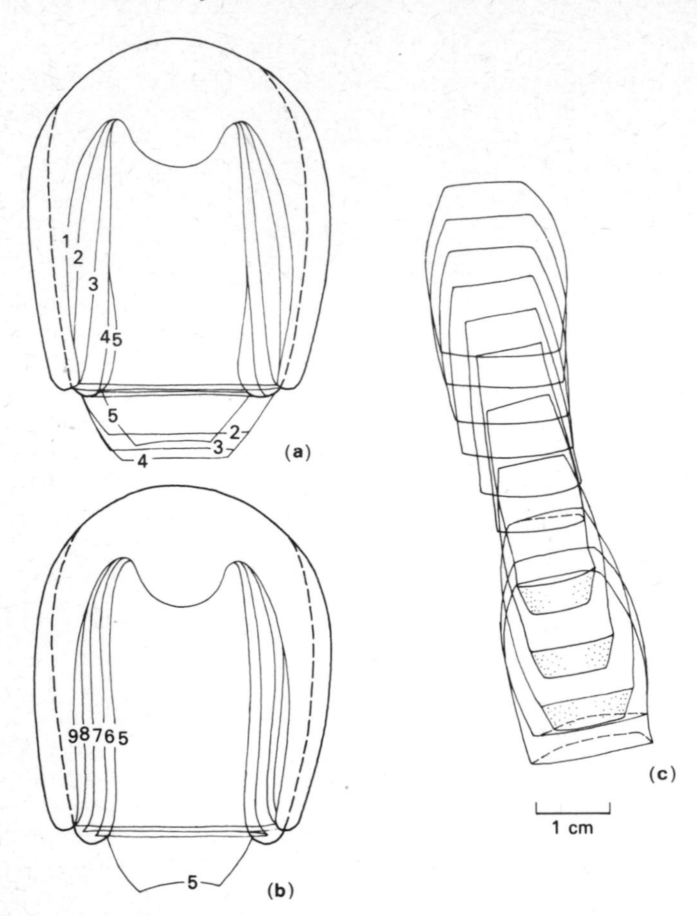

Fig. 6.16 Contraction (**a**) and recovery (**b**) profiles of *Polyorchis* during a single cycle traced from cine film, numbers indicate successive stages at time intervals of $\frac{1}{12}$th s; (**c**) lateral aspect of sequence showing change in position at intervals of $\frac{1}{16}$th s. (from Gladfelter[39])

the bell with the folding of the mesoglea around a series of eight structural joints in the adradial regions.

Successive jet cycles have been filmed and it is possible to estimate jet thrust and pressure by use of Gladfelter's analysis and the equation (2, p. 133) relating jet thrust and pressure. A large specimen of *Polyorchis* (25 g) develops a pulse of 0·33 s duration and expels 6 g of water through a velar aperture of 1·35 cm^2 at the middle of the contraction phase. This results in an increase in velocity of swimming of $2 \cdot 3 \text{ cm s}^{-1}$ and theoretically requires a pressure pulse of $1 \cdot 2 \text{ N cm}^{-2}$ amplitude if the coefficient of discharge is 1. With a lower efficiency of discharge even greater pressures would have to be produced. Unfortunately we have no pressure traces from *Polyorchis* but recordings made of swimming by a disc-shaped scyphozoan, *Cassiopeia*, gave pressure pulses of only $2 \cdot 5 \times 10^{-3} \text{ N cm}^{-2}$. Analysis of a cine film taken of the swimming of this jellyfish of $3 \cdot 5$ cm diameter using the same method as for *Polyorchis* gave a value for pressure of only $5 \times 10^{-3} \text{ N cm}^{-2}$. This latter was, however, difficult to calculate accurately since the contractions of the disc are limited to a narrow peripheral band and so must be regarded only as an approximation. Nevertheless, there is a considerable difference between *Polyorchis* and *Cassiopeia*, and the pressures determined suggest that bell-shaped jellyfish are powerful swimmers with a high pressure pulse and are shaped to reduce the effect of drag. The disc shape, by contrast, increases drag and prevents the animal from falling too readily through the water, its position being maintained by a series of low pressure pulses. Clearly more experimental work is required to elucidate this problem of the relationship between shape of jellyfish and the dynamics of their locomotion.

CONCLUSIONS

The effectiveness of jet propulsion as a mechanism both for fast and for sustained swimming has been discussed. Undoubtedly *Loligo vulgaris* is one of the best examples, for this squid has slight negative buoyancy and must swim to maintain its position. This is achieved by means of deep respiratory pressure pulses which serve both to ventilate the mantle cavity and for locomotory purposes. The same circular muscles of the mantle generate high pressure pulses, giving the squid an acceleration of more than 3 *g*. This occurs during escape swimming and cannot be long sustained. In other animals, such as the scallop, escape from predators must also have been an important factor in the evolution of jet swimming.

All animals using this method of locomotion must eject as large a volume of water as possible; thus during ontogeny the rapid development of the largest possible mantle or subumbrella cavity is at a premium. For maximum velocity the animal must not only be streamlined but must be capable of ejecting the fluid at high pressure. This requires the development of substantial musculature.

The final problem encountered in jet swimming is that of refilling the water-filled cavity and stretching the propellor muscles. This is carried out by the elastic properties of the mesoglea in jellyfish, and of the hinge ligament in scallops, serving as potential energy stores, and by the antagonism of radial mantle muscles in squid. Apart from normal swimming in scallops, none of these examples allows the refilling of the cavity to occur by 'ram' pressure as they move through the water. The influx of fluid by utilization of a small part of the energy of propulsion would seem a natural adaptation for jet propulsion but it is only salps that seem to be ideally designed for this purpose. Salps are tubular in shape, having circular muscle fibres which supply the propulsive force.[13] The presence of valvular mechanisms at the inhalant (anterior) and exhalant openings suggests that the inhalant aperture will be closed as water is expelled and that the resulting forward motion may be accompanied by closure of the exhalant siphon as the inhalant opens. This would allow the branchial cavity to be filled and the circular muscles to be stretched by the inflow of water as the salp moves forwards. In addition, the semi-rigid test of a salp is deformed by contraction of the muscle bands and the original body shape is regained by its store of potential energy[17] in a similar manner to the refilling of the branchial cavity of a sessile ascidian (p. 14). In both examples, loss of water by jetting is incompatible with the use of the branchial cavity as a fluid skeleton for muscular antagonist.

The means of locomotion of a salp appears to be basically so simple and potentially so successful that one wonders why the tubular shape with valves at each end has not been more widely evolved for jet propulsion. The answer may lie in the manner in which animals have adapted their basic body form. They have also to feed and to reproduce, and it would be difficult to envisage this plan superimposed on the morphology of a squid or jellyfish. But full discussion must await detailed experimental analysis of the locomotion of salps.

7

Locomotion and Metazoan Evolution

INTRODUCTION

So far in this book we have been largely concerned with the mechanisms of locomotion of soft-bodied animals in a variety of habitats. The conclusions will now be related to animal phylogeny, with particular reference to the ways in which the requirements of locomotion have been of prime importance in determining the course of evolution. However, other problems, such as the origin of the Metazoa, fundamental to the understanding of the patterns of metazoan evolution, must first be briefly considered. The introduction of new locomotory techniques, as evolution proceeds, will be shown to be associated with the rapid exploitation of some new habitat with consequential changes in structure and organization of the organisms. Indeed it is difficult to envisage any other single factor than locomotion having such profound effects on the development of organ systems in Metazoa.

The analyses of locomotory mechanisms of soft-bodied animals, presented here, indicate how locomotory habit and body structure may be related. In general there seem to be comparatively few ways in which the demands of locomotion can be met, so that animals of widely different phylogenetic origin may show similar adaptations in response to similar mechanical problems. This is well exemplified in respect of burrowing through sand or mud, where the principal of anchoring part of the body whilst another part moves forward is developed in all major phyla of soft-bodied animals. This results in the alternate use of two types of anchorage, penetration and terminal (p. 43). Different fluid-filled body cavities have evolved to provide a mechanism allowing either extension or distension of the body wall. This occurs by use of the coelenteron in anemones, for example *Peachia*, the coelom in worms, for example *Arenicola*, or the haemocoel in molluscs, for example *Ensis*. Further, the mechanisms used for initial entry into the sand by *Peachia* and *Arenicola* are very similar (p. 48). These animals of different phyla exhibit convergent evolution in respect of the mechanical problems encountered in soft substrates. Each example mentioned above is circular in cross-section and can change the shape of a significant part of the body so as to become anchored or extend through the sand, but nevertheless each uses a different type of body cavity as the basis of their hydraulic system.

The same principal of anchorage and extension obtains in respect of hard-bodied animals but it is obscured by the use of multiple appendages such as the limbs of crustaceans or the spines of echinoids. At any instant some appendages would be extending further into the sand while others give anchorage by thrusting laterally or backwards.[121] Thus in these two groups we find locomotory adaptations different from those of worms or molluscs. Exceptions of this sort are readily understood and are met relatively infrequently, for similar locomotory requirements in animals of the same grade of organization generally result in the evolution of similar structures. We may thus have confidence in any conclusions relating to movement and habit drawn from examination of locomotory structures.

It is important to realize that any evolutionary deductions, based on observations and experiments with living animals, must apply equally well to extinct animals since the demands of locomotion are essentially physical and mechanical and are accordingly predictable. Thus deductions can be made about the probable habitat of a fossil from an examination of its potential locomotory powers.

Much of what has been written on problems such as the origin of the Metazoa, or the inter-relationship of invertebrate phyla, could be considered to be in the realm of fantasy.[54] This is because so much of our evidence is indirect, being based on comparisons of structure and development of present day animals or deductions from a restricted range of fossils. What is strikingly in evidence is the rapid evolution of most major animal phyla at approximately the beginning of the Cambrian period and their consequent adaptive radiation. Details of this process are necessarily obscure and by the very nature of the evidence universal acceptance of schemes for the derivation of major phyla will long be in dispute. Nevertheless, phylogenetic speculation will continue and must be restrained by some rules; for example, all hypothetical ancestral forms must be viable living animals, and any innovation thought to occur must confer some selective advantage. It is in this context that studies of locomotion, based on the realism of experiment and observation, make an ideal approach to phylogenetic studies.

AN OUTLINE OF EARLY METAZOAN EVOLUTION

Origin of the Metazoa

After a long history of evolution of protozoan-like organisms, multicellular animals and plants arose and radiated suddenly near the beginning of the Cambrian period. Palaeontologists have attempted to explain these sudden developments in terms of either physical and chemical controlling factors such as the levels of ultraviolet radiation and oxygen in the atmosphere, or major adaptive breakthroughs such as the appearance of shell or cuticular secretion.[97] Recently an ecological theory has been suggested by Stanley[103] based on the premise that high diversity at any trophic level only occurs under the influence of cropping. Until herbivores evolved, the protistan algae of the Precambrian were limited by resources rather than by predation and large populations of a small number of species probably occurred so as to saturate the aquatic environments. In the near absence of vacant ecological niches life would diversify only slowly. There is good evidence of planktonic autotrophs being established more than 1300 million years ago, while pelagic heterotrophic protists probably first occured some 900 to 700 million years ago.[97] The latter possibly fed first on bacteria, subsequently on algae, as the ability to ingest larger particles of food evolved. From the consumption of algae, little modification would be required to enable these heterotrophs to become carnivorous in habit. Thus Stanley considers that the adaptive breakthrough to algal feeding led rapidly to the addition of diverse trophic levels and the proliferation of feeding networks by adaptation to the vast variety of ecological niches then present both in the planktonic and benthic habitats. In this rapid diversification of life over a period of probably no more than a few 10 millions of years the Metazoa and Metaphyta arose almost simultaneously.

The early metazoans were soft-bodied and probably pelagic, initially using phytoplankton as food. Later they found the ocean floor and were able to feed on growths of encrusting algae. It was in these circumstances, at the very end of the Precambrian period, that bilateral symmetry probably first occurred as a consequence of locomotion over substrates. At about the same time hard skeletons may have been first formed as a response to predation pressure, giving rise to the diverse fossil record of the Cambrian and the succeeding period, the Ordovician. Thus in a relatively short time fossil remains of the major phyla suddenly appear.

There are two main theories to account for the origin of the Metazoa. The first, based on the integration of protistan colonies

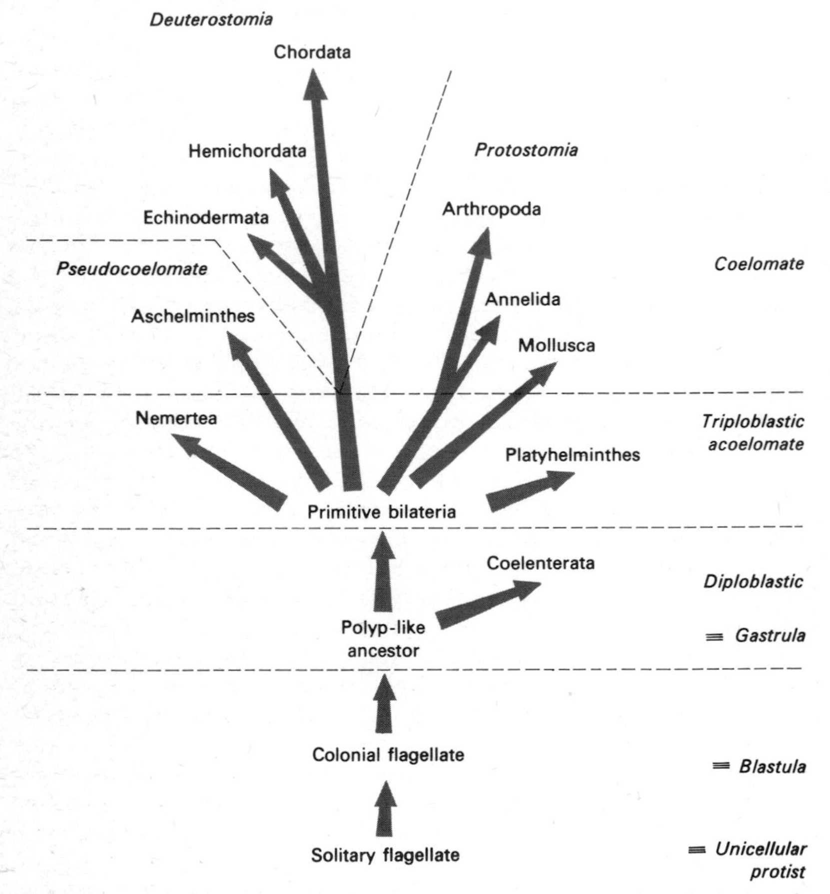

Fig. 7.1 Generalized diagram indicating possible relationships between the principal metazoan phyla on the basis of the gastraea theory. Different grades of organization are also indicated.

and suggested by Haeckel in 1874, dominated zoological thought on this subject for the first half of the present century (Fig. 7.1). It presented the Cnidaria, as exemplified by *Hydra*, as the most primitive metazoan phylum, originating from colonial protists in the form of a hollow ball of cells (blastaea) rather similar to the colonial flagellate, *Volvox*. The crucial evolutionary step was the separation of locomotor and nutitive functions and the invagination of the cells concerned with nutrition, so that the cavity within the animal was bounded by a wall containing two cell layers (diploblastic). This gastraea stage was thought to be ancestral to all existing Metazoa and to be still represented in the ontogeny of many animals by the gastrula. On the basis of Haeckel's gastraea theory, the Porifera, Cnidaria and Ctenophora all remain at the gastrea level in organization while evolution of the higher animals is envisaged by the addition of a third cell layer, the mesoderm, to some simple polyp-like form.

The gastraea theory of the origin of the Metazoa is based on the dogma of recapitulation, that adult forms succeed each other and illustrate their phylogeny in their developmental stages. Today it is apparent that the relationship between ontogeny and phylogeny is complex. It is realized, for example, that animals have to be well adapted to environmental conditions at all stages of their development and that when neoteny occurs the adult descendant resembles the juvenile stage of an ancestor, the reverse of what might be expected under the recapitulatory hypothesis. Other weaknesses of the gastraea theory are the uncertainty as to whether the Cnidaria are composed of only two cell layers, and that the manner of gastrulation in Cnidaria is usually by cell migration rather than by invagination. Clark[26] further points out that some modern zoologists supporting this theory, for example Jägersten and Marcus, couple it with an enterocoelous manner of origin of the coelom (p. 170) from four coelenteric pockets occurring in the gastraea stage. This leads to the view that the most primitive triploblastic animal was coelomate and segmented, all acoelomate and unsegmented animals being therefore degenerate. This appears to be an improbable interpretation.

The gastraea theory regards the Protozoa as essentially unicellular organisms which came together to form a metazoan by a process of aggregation. Thus each protozoan is thought to be equivalent to one cell of the many comprizing a body of a metazoan. A single protozoan, however, performs all the functions carried out by a metazoan and could be equated with the whole metazoan body. On this basis the Protozoa should be thought of as acellular animals and the Metazoa to have originated from multinucleate or syncitial protozoans, similar to some ciliates, by the development of cellular boundaries. This theory of origin by the process of cellularization was initially based on speculations of zoologists such as Ihering, Saville-Kent and Sedgwick, during the latter part of the nineteenth century, but it has been recently revived by the writings of Hadži.[46] Hadži considers that an acoel turbellarian, such as *Convoluta*, is comparable to many ciliates, in being ciliated, of similar size and lacking a hollow digestive cavity, the central parenchymatous tissue of Acoela being imperfectly cellularized (Fig. 7.3b). This hypothesis places the Acoela at the base of the Metazoa in place of the cnidarian polyp form, so raising the question of whether the Cnidaria are to be regarded as primitive (Fig. 7.2). In the syncitial protistan theory both Cnidaria and Ctenophora must be considered as reduced triploblastic animals, probably derived from rhabdocoel turbellarians. This involves the reversal of the conventional view of the Cnidaria, by interpreting the Anthozoa as the most primitive members, and the structural simplicity of the Hydrozoa as a

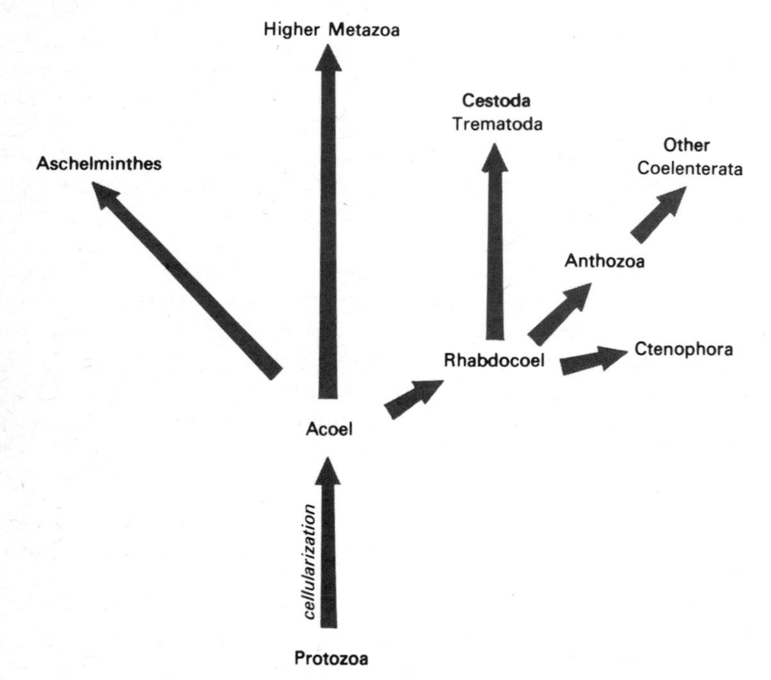

Fig. 7.2 Diagram of the phylogeny of the lower Metazoa according to Hadži's syncitial protistan theory.

secondary development. This view is strongly opposed by authorities on the coelenterates, such as Hyman[54] and Pantin.[90]

One major objection to the gastraea theory is that in most primitive living metazoans the endoderm is formed, not by invagination, but by ectodermal cells wandering into the centre of the blastula. The larva thus resembles a planula larva of cnidarians with a solid cellular interior, rather than a gastrula with a large hollow digestive sac (Fig. 7.3). While there would be considerable advantage

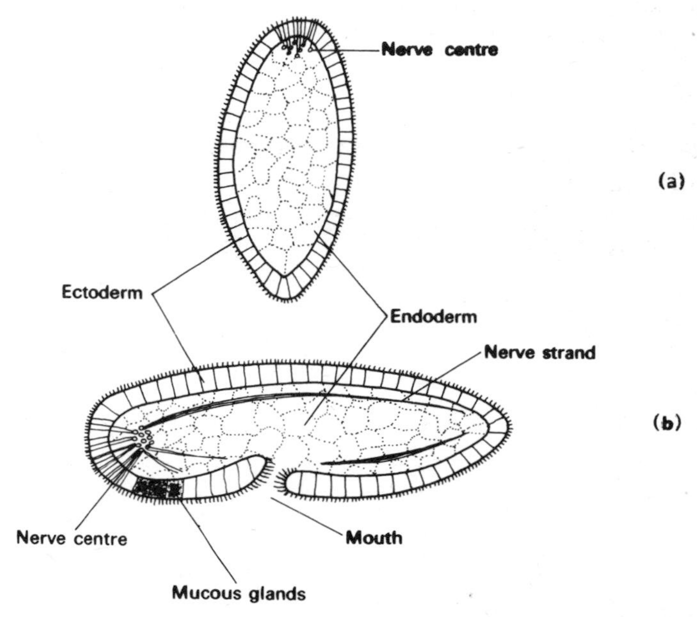

Fig. 7.3 Diagrams illustrating planuloid-acoel theory of the origin of the Bilateria. (**a**) Planula larva, mouthless and with solid endoderm; (**b**) acoel stage, a bilaterally symmetrical worm creeping by means of cilia with the mouth leading to interior mass of nutritive cells (after Hyman[53]).

to the early metazoan in the separation of locomotory and sensory functions from the digestion of food, it does not follow that a digestive sac is immediately required. Intracellular digestion, widespread in the lower Metazoa, would render the formation of a mouth and digestive sac superfluous and it would be reasonable to think of the most primitive metazoan being a solid organism similar to the planula larva of present day Cnidaria. This planuloid–acoel theory, supported by Hyman[54] and Hand,[49] allows the descent of both the coelenterates and the Turbellaria from planulae, retains the coelenterates as the most primitive metazoans, albeit as a side branch of the main evolutionary line, and allows the Platyhelminthes, Nemertea, Aschelminthes and the higher metazoa to be evolved from an acoel stem form (Fig. 7.4). The problem of whether

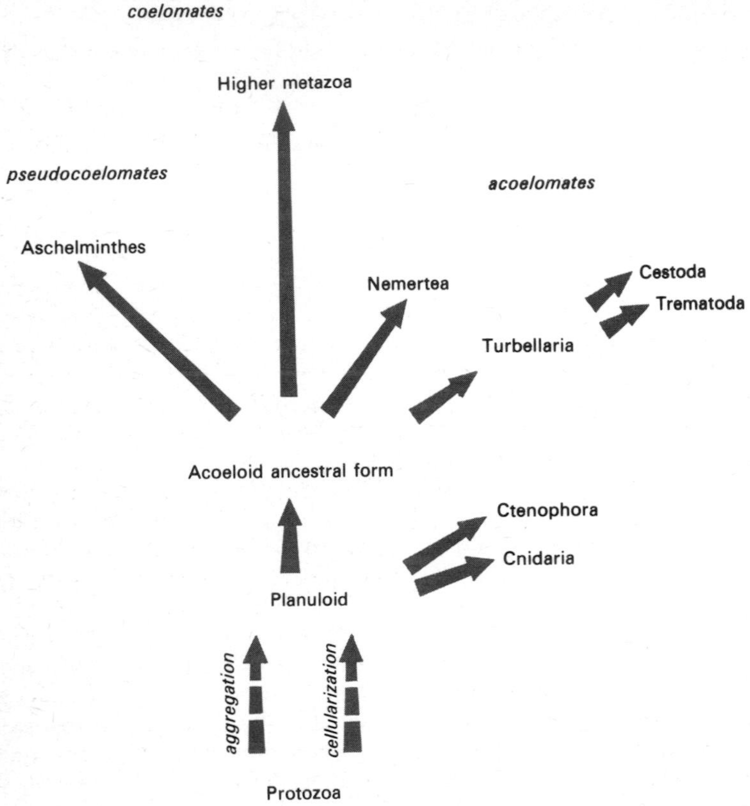

Fig. 7.4 Diagram of the possible phylogeny of the lower Metazoa based on the planuloid-acoel origin of the Bilateria.

a planuloid larva arose from a colonial protistan by inward migration of specialized absorptive cells or by cellularization of a multinucleate protozoan is of only theoretical importance to the arguments concerning locomotion and the evolution of the metazoa.

The major evolutionary step from planuloid to polyp-like animal is the development of a coelenteron. This may have arisen primarily to attain a large surface area for food ingestion and then have subsequently acquired the role of hydrostatic skeleton. An alternative possibility that the fluid skeleton evolved first in response to a burrowing habit should not, however, be overlooked.[117]

Early evolution of bilaterally symmetrical animals

Some early systems of classification of the animal kingdom distinguished between bilaterally and radially symmetrical animals, the latter being a miscellaneous collection of phyla including the coelenterates. Subsequently Hatschek[51] used the term Bilateria to designate all the Metazoa above the Coelenterata and despite some modifications the term has continued in use, for with few exceptions all coelomate and segmented animals can be shown to be bilaterally symmetrical.

Whatever the manner of origin of the Metazoa, the earliest representatives must have consisted of relatively few cells and have been comparable in size to a large ciliate or small acoel with a length of about 1 mm. They were likely to have been covered in cilia and to have swum around as planktonic organisms. Any increase in size would probably have resulted in descent to the sea bottom, where they could creep by means of the cilia of the ventral surface in the manner of free-living flatworms. The surface of the substrate would be rich in food, receiving a rain of detritus from the ocean above, and would present a habitat likely to be exploited by an early metazoan. Such an organism would have few contractile fibres and, without these, neither could sufficient force be exerted nor could there be changes in shape adequate to permit adoption of the burrowing habit as an alternative way of life. These early metazoans could, however, have exploited the interstitial habitat as members of the meiofauna by remaining small and becoming elongated in shape. This mode of life, moving between sand grains without disturbing them, has occurred in a variety of phyla, for example the small hydroid *Psammohydra nanna*, the archiannelid *Diurodrilus minimus* and the holothurian *Leptosynapta minuta*,[105] but life in this habitat is highly specialized. It is reasonable to take Clark's[26] view that the early Bilateria occurred at or just above the substrate. Increasing size would lead to greater nutritional requirements, and life on the bottom to feed on organic detritus is the habit most likely to have been exploited.

A consequence would be the development of a mouth and digestive system. An initial inability to select food may have led to mass ingestion with the consequent need of a relatively extensive gut to deal with the problems of intracellular digestion. Undoubtedly such an animal would have first moved by cilia, requiring the secretion of a mucous bed in which to beat in the manner already described for planarians (Fig. 2.1b). An early metazoan may well have been radially symmetrical when planktonic, perhaps as a planula-like organism, but adoption of a creeping habit on the sea bottom must have resulted in the rudiments of bilateral symmetry being developed in respect of anterior mucous glands and a sensory system (Fig. 7.3). This symmetry subsequently became fully established with the development of muscles for more powerful locomotion and a rudimentary nervous system for their control.

With increased size the animal would tend to become flattened dorso-ventrally for the purposes of respiration and to ensure that there was a proportionate increase in the area of locomotory apparatus bearing on the substrate. Such a shape has become fixed in free-living Platyhelminthes with the development of dorso-ventral musculature, while longitudinal muscles also arise and show a tendency to replace cilia in locomotion. Longitudinal muscles may contract to cause an overall shortening of the body as in looping motion, or to cause local shortening and thickening of the tissues. Each of these thickened regions could thus form a *point d'appui* and, provided relaxation and contraction of the muscle occurs cyclically, retrograde locomotion (Fig. 2.7). Retrograde, rather than direct, locomotory waves occur in Turbellaria, Nemertea and in the feet of many molluscs, including chitons and the more primitive gastropods (Prosobranchia). It is thus reasonable to assume that retrograde locomotory waves are more primitive than direct waves. Further supporting evidence comes from consideration of how a flatworm moves forward by muscular activity. The head is first raised and then extended by the tissues becoming thinner by contraction of dorso-ventral muscles whilst anchorage is obtained posteriorly by local contraction of longitudinal muscles. This process is essentially the same as occurs in retrograde locomotion in the limpet (p. 31). By contrast an animal using direct waves for locomotion, such as a terrestrial pulmonate, must first raise the tail and pull it forwards (p. 37). The forward thrusting of the head associated with retrograde waves is a much more likely starting point for muscular propulsion than retraction of the tail.

The amount of forward motion of the head of a flatworm must necessarily be restricted by the limitations of dorso-ventral muscles and the parenchymatous nature of the tissues. This form of tissue probably suffices for localized muscular antagonism as it has been shown to occur in the cephalopod mantle (p. 148). However, where muscular forces need to be transferred over relatively large distances, or where major change in body shape is required, then large fluid-filled cavities are found as, for example, in burrowing worms or molluscs. Fluid-muscle systems have been classically described in respect of animals containing a relatively large volume of free liquid in the coelenteron as in *Metridium*, the haemocoel as in *Ensis*, or in the coelom as in the earthworm and *Arenicola*.[22] This emphasis was probably a result of these animals being the clearest examples of the hydrostatic skeleton. However, the importance of body tissues in respect of local force transference should not be overlooked. These tissues suffice for movement over firm substrates both in Turbellaria and Mollusca, but whenever penetration of sand or mud is achieved body cavities containing relatively large amounts of fluid are essential both for force transference and dilation of part of the body wall.

The evolution of the coelomate condition

The sea bed presented the primitive Bilateria with an environment rich in food and small microphagous and detritus feeders must have been amongst the first inhabitants. As soon as sufficiently large populations were established, these animals would be followed by specialist carnivores. Shallow burrowing was likely to have occurred at this stage as an escape from predation while remaining in contact with the rich food supplies at the surface.

A fundamental change in shape from flattened to circular cross-section follows the adoption of an infaunal habit. An animal of circular cross-section is ill suited to surface locomotion because only a small part of the body wall is in contact with the ground, but it is well suited for burrowing, since it is everywhere in contact with the substrate and all the musculature is able to contribute to locomotion. A circular cross-section allows the development of circular muscles to antagonize the longitudinal fibres, for instance in peristaltic movements as in *Cerebratulus* (p. 24), but this situation cannot be exploited until a true fluid skeleton is evolved.

In planarians the parenchyma is relatively closely packed and allows little change in body shape, although permitting local muscular antagonism. The nemertines are of a similar grade of tissue organization to flatworms, but show a reduction in the number of parenchyma cells and an increase in interstitial fluid. This quasi-fluid skeleton enables nemertines to undergo larger changes in shape with greater powers of muscular movement and some are able to burrow.[121] Nemertea additionally possess a true fluid filled cavity, the rhynchocoel, which allows the proboscis to be everted powerfully. The proboscis is used in *Cerebratulus* both to penetrate the sand and as a terminal anchor. Nevertheless the rhynchocoel does not provide a satisfactory alternative to a fluid skeleton for the antagonism of the body muscles, and the nemertines have met with only partial success as infaunal animals.

The increase in interstitial fluid within the parenchymatous tissue has only limited success in solving the problem of a fluid skeleton, whereas the evolution of a secondary body cavity in the form of a coelom has been very successful in a wide variety of animals. The coelom has a great potential advantage as the basis of a hydraulic system for an infaunal life and once achieved in some primitive bilaterian has led immediately to radiation (Figs. 7.5 and 7.6). It is likely to have arisen on several occasions during evolution. Evidence for this lies in the two distinct embryological origins of the coelom in invertebrates. Firstly, the coelom may arise by a split occurring in the mesoderm, forming what is termed a schizocoel. Most animals in which this occurs are members of the Protostomia, for example Annelida, Arthropoda, Mollusca. Secondly, it arises from cavities in mesodermal sacs which evaginate from the archenteron. This is termed enterocoelic development and is commonly seen in the

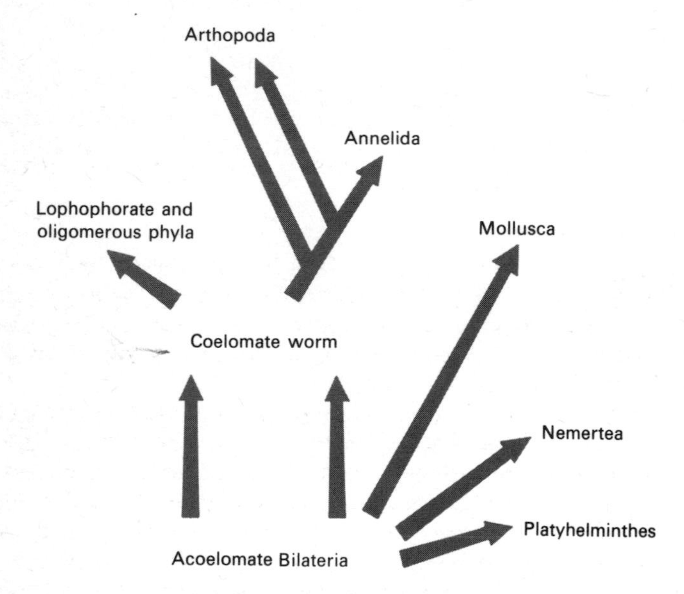

Fig. 7.5 Diagram indicating possible relationships between the principal protostome groups. Repetition of arrows suggests repeated origin.

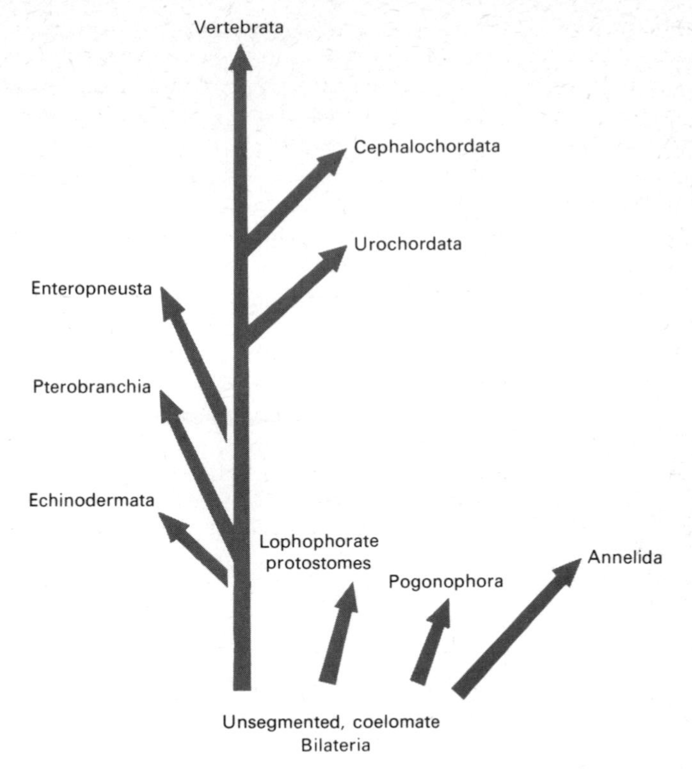

Fig. 7.6 Diagram of possible relationship of deuterostomes, pogonophores and lophophorate protostomes.

Deuterostomia as represented by the Echinodermata and the chordate line of evolution.[53] This difference in origin of the coelom is not put forward here as a phylogenetic argument to differentiate between the Protostomia and Deuterostomia but to indicate two modes of development of the coelom.

There are numerous theories of the origin and evolution of the coelom which are critically summarized by Clark.[26] These include the gonocoel theory, which is widely held and is based on the common association of gonads and coelomic epithelium, the coelom being regarded as the cavity of an expanded gonad; the enterocoel theory suggests that the embryological origin from gastric pouches is primitive, possibly occurring by separation from the main digestive cavity in certain Anthozoa; the nephrocoel theory proposes that the coelom originated as an expanded nephridium; and the schizocoel theory regards the coelom as a new formation in the mesenchyme unrelated to gonads or gut and without antecedents in the lower Metazoa. Although there is evidence, largely embryological, which may be cited to support any of these possibilities, there is nothing conclusive and the method of evolution of the coelom must remain in doubt. Some theories are more suited to explain the origin of the coelom in certain phyla, for example the gonocoel theory in respect of molluscs or the enterocoel theory for echinoderms. The view that the coelom arose not once but in different ways in different phyla, by what was effectively convergent evolution, is the most reasonable conclusion.[25] But how the coelom has originated is less important than recognition of its function as a hydraulic system.

The coelom provided the early Bilateria with the means by which they were able to burrow, probably shallowly at first, feeding on surface detritus by various devices such as a proboscis, mucous net or ciliary filtration system. Many modern sedentary worms living in this way pass a water current, for respiratory purposes, through the burrow by peristaltic contractions of the body wall, using the same musculature and fluid skeleton as has enabled burrowing to take place. Co-ordination of the muscles of an elongate body is now associated with both locomotion and respiration, and leads to further development of the nervous system.

The evolution of a true hydrostatic skeleton not only enabled the early Bilateria to burrow by making powerful movements but also had important secondary consequences. Perhaps most important amongst these was the intimate association of gonads and coelom. The large coelomic cavity allowed the storage of gametes without impairing locomotion, so that the breeding season could be synchronized throughout a population, resulting in a brief but intensive breeding period. This would both increase the chances of fertilization and ensure that the young are produced in optimum conditions. Clark[26] points out how hormones, secreted from neurosecretory cells, probably effect the synchrony of gonadial maturation and suggests that the presence of a secondary body cavity demanded a hormonal control system. The coelom indeed provides the fundamental reason for the evolution of an endocrine system. The coelomic fluid bathes all organs and has an important role in transportation of metabolites and possibly oxygen between the body wall and the digestive system. In annelid worms the coelomic cavity is separated from the epidermis by relatively thick muscle layers and, with increasing size, a blood vascular system must have made its appearance to provide a transport system from the epidermis or special gill structures to the more centrally located organs. Finally the coelomic cavity allows further development of the excretory system. In a flatworm this consists of numerous protonephridial tubules located throughout the body within a short distance of all tissues. In a coelomate worm excretory products may diffuse into the coelomic cavity whence they may be removed by a relatively small number of more highly developed nephridia. These secondary consequences have led to the great biological success of coelomates and far exceed the advantages gained initially by use of the coelom as a hydraulic system in burrowing.

The pseudocoelomates

Pseudocoelomate animals include the Nematoda, Rotifera, Gastrotricha, Kinorhyncha, Acanthocephala and Endoprocta. All, with the possible exception of the last two groups, are regarded as members of the phylum Aschelminthes but their relationships are obscure and the phylum is best considered to be polyphyletic in origin. These animals have a common characteristic in that the body cavity is in the form of a pseudocoel, a cavity derived directly from the embryonic blastocoel. The pseudocoel is effectively a fluid-filled cavity between body wall and gut, somewhat comparable to the coelom, although of fundamentally different origin. The best example of the use of the pseudocoel in locomotion is *Ascaris* which has been discussed in Chapter 5. How far the mechanical analyses of the form and movement of nematodes can be applied to other pseudocoelomates is uncertain, for they are all small in size and there are few observations on the role of the pseudocoel in locomotion. One exception is the Acanthocephala where Hammond[47] observed two hydraulic systems, one concerned with the protrusion of the proboscis and the other with movements of the trunk.

It seems likely that the blastocoel has persisted in these animals as an alternative form of body cavity to the coelom. This would only occur on the basis of some functional requirement, such as the need for powerful movements, or because of neoteny taking place in the early evolution of these groups. In view of their small size the former is unlikely; but a neotenous larval form might have a body cavity of potential significance for the radiation of this group (Fig. 7.4). The advantages of hormonal mechanisms has already been mentioned and it is likely that, at an early stage of the evolution of the Bilateria, hormones controlled growth and reproduction. Hormones may have functioned relatively imperfectly at first, synchrony of somatic and sexual development being but poorly controlled and accordingly increasing the chances of neoteny occurring. In these circumstances, persistence of the blastocoel as an alternative form of body cavity to the coelom in the early Bilateria would hardly be surprising.

The evolution of metameric segmentation

Metameric segmentation of animals involves the division of the body transversely into a series of sections each essentially having one pair of the various organ units of mesodermal origin. Such an arrangement is an ideal condition, for in many animals it has permitted the economical restriction of specific organs to a limited number of segments, as, for example, the reproductive organs in an earthworm.

We have seen how the coelom probably originated in the early Bilateria as a hydrostatic skeleton in response to the adoption of a burrowing habit. One of the disadvantages of a single coelomic compartment is that muscular activity in any part of the body affects the whole fluid–muscle system. Metamerism involving the coelom, as in annelids, allows localized muscular activity to take place; in the earthworm, for instance, pressure varies between adjacent segments during locomotion.[98] The segmental organization of the hydraulic system does not allow coelomate worms to carry out new types of movement, but it improves their efficiency by facilitating localized changes in shape, so ultimately freeing them from a sedentary existence.

Clark[26] suggests that compartmentalization of the coelom first occurred in coelomate worms while they were still relatively unspecialized and he rejects the possibility of oligomerous animals, similar to present-day lophophorates or pterobranchs, being ancestral to septate worms, on the grounds that their tentacular condition would introduce an unnecessary complication. Oligomerous animals characteristically have the coelom divided into first a protocoel functioning as a skeletal system for the proboscis, secondly a mesocoel supporting feeding tentacles and thirdly a metacoel providing the trunk coelom. The development of feeding tentacles and provision for their support in a coelomate worm suggests a habit of feeding on particles in suspension. This is unlikely to have occurred in the evolution of metamerism in worms and to have been lost in oligochaetes and errant polychaetes. Nevertheless, metamerism may well have originated by compartments forming in the anterior part of a coelomate worm in response to a need for greater local control in burrowing, particularly in respect of the development of a specialized tool for the excavation of a burrow in the form of a proboscis. In *Sipunculus* and *Priapulus*, both of which are coelomate and primitively unsegmented worms of burrowing habit, movement is brought about by the forcible extension of the proboscis with constriction of the posterior trunk region and the generation of a high pressure pulse. The consequent loss of the penetration anchor is incompatible with powerful forward thrusting through the substrate (Fig. 3.8). Both of these animals obtain additional purchase on the walls of the burrow by flexure of the body.[48] Such forceful eversion of the proboscis, unspecialized for scraping through sand, is also particularly unsuited for initial entry into the substrate (p. 55). Division of the coelom allows regions of the body to act independently of the whole and nowhere can this have been more important than in the head of a burrowing worm, for this allows the development of the proboscis as an excavating tool independent of the main trunk coelom. *Arenicola*, with a septate anterior region of the trunk (Fig. 3.6), has secondarily achieved this condition by regression from the septate condition. It possesses a proboscis which can excavate the burrow at low coelomic pressures (p. 49). *Leptosynapta*, a burrowing holothurian, uses tentacles to excavate the burrow, and peristalsis of the trunk body wall to progress through the burrow.[35] The trunk is supported by the main perivisceral coelom while the hydrocoel is present in the tentacles so that these may act independently. Division of the coelom anteriorly by septa, for functional reasons concerned with an infaunal life, could lead by repetition of the septa to segmentation as in annelids and could also be envisaged as a starting point for the tentacular oligomerous phyla (Fig. 7.5).

In annelids the septate condition led to the specialization of muscles and nervous system on a segmental basis so that each segment could act as a nearly autonomous unit, but with its activities co-ordinated with the remainder of the body. Lack of intersegmental flow limits the use of the coelomic fluid as a means of transport of oxygen or metabolites, so that an effective vascular supply for each segment becomes essential. Similarly, the excretory and reproductive systems acquired a segmental pattern because of a restriction on intersegmental movement.

Chordates also exhibit metameric segmentation but this has arisen independently of the annelids in association with the evolution of the notochord and swimming by undulatory body movements. In both groups the undulatory waves are produced by flexure of the body, in chordates by the application of torsional forces to the axial skeleton, but with a fluid skeleton, as in worms, the only force that can be applied is compressional. The longitudinal muscles of one side of the body contract and antagonize those of the other side by means of fluid pressure, so causing the body to be flexed. This movement is carried out equally well whether or not the worm is segmented (p. 124). Thus we may conclude that in annelids segmentation occurred to give greater mechanical efficiency in burrowing, but that in chordates it evolved in respect of undulatory swimming. Another type of metamerism, in the sense of the serial repetition of body organs, occurs in the cestodes, principally for reproductive needs. Thus serial repetition has evolved at least three times in the animal kingdom, as response to three different evolutionary demands.[25]

PROTOSTOMIA AND DEUTEROSTOMIA

The higher Metazoa may be divided into two main lines of descent, the Protostomia and Deuterostomia, on the basis of embryological criteria, for example the fate of the blastopore as mouth or anus respectively, spiral or radial cleavage of the egg and schizocoelous or enterocoelous formation of the coelom. Associated with such embryological features, the larva is generally characteristic as a trochophore or dipleurula larva respectively. The protostomes include such phyla as the Annelida, Arthropoda and Mollusca while the principal members of the deuterostomes are the Echinodermata and Chordata.

Three other protostomatous phyla also merit consideration because of apparent deuterostome affinities. These are the Phoronida, Ectoprocta and Brachiopoda, collectively termed the lophophorates, characterized by the presence of a horseshoe shaped food-catching organ, the lophophore, bearing numerous ciliated tentacles each containing an extension of the coelom. Many deuterostomes are oligomerous, such as members of the Echinodermata and the Hemichordata, the coelom and often the body also being divided into three regions, at least in the larval stage. These lophophorate and deuterostome phyla probably evolved from primitive burrowing coelomates to exploit a different sort of food. The early coelomates exploited the rich detrital deposit on the sea bed while gaining protective advantage from an infaunal life. The oligomerous animals probably also gained protection by burrowing but developed a tentacular apparatus to collect detritus as it settled. For this to be possible at least two, probably three, coelomic compartments are required, each associated with a particular region of the body. From the anterior end these would be first, a hydraulic organ (protocoel), to facilitate penetration of the substrate somewhat similarly to the polychaete proboscis; secondly, a cavity (mesocoel) for tentacular support and protrusion; and thirdly, a trunk coelom (metacoel) to effect burrowing by peristaltic movements of the body wall. Such an organization is similar to that of the Pterobranchia (Phylum Hemichordata). However, in the lophophorate phyla there is no well developed protocoel and it is unlikely that the phoronids, for instance, ever possessed a protosome.[54] The lophophore is carried on the mesosome while the trunk represents the metasome in adult *Phoronis*, which thus has a dual hydraulic system separating the functions of tentacular support from burrowing. Such an arrangement resembles that found in some sedentary polychaetes, for example *Terebella*, in which only a single septum remains, isolating the tentacular region from the trunk.

Differences in the detailed organization of oligomerous animals makes it likely that they are a polyphyletic assemblage and that the coelomic arrangement represents a grade of construction rather than a characteristic of a natural group. Nevertheless their embryology is relevant to discussion of the origins of the protostomes and deuterostomes. The echinoderms and hemichordates are clearly members of the latter assemblage, for the mouth forms a secondary opening at the opposite end of the embryo from the blastopore, but lophophores display features belonging to both groups. In phoronids and brachiopods the mouth is derived from the blastopore and the larvae of the phoronids (actinotroch) and the ectoprocts (cyphonautes) are comparable to the annelidan trochophore. These are clearly protostome features yet cleavage is radial in both ectoprocts and brachiopods and in some members of the latter phylum the coelom arises enterocoelically as is typical of deuterostomes. The combination of protostome and deuterostome features in lophophorate phyla and the oligomerous condition of hemichordates suggests that the point of origin of the lophophorates may be near that of the deuterostomes (Fig. 7.6).

The phylum Pogonophora represents another vermiform group whose body is divided into three regions, of which the anterior cephalic lobe bears tentacles (Fig. 7.7). It has been customary to assign them to the Deuterostomia with close affinities to the Hemichordata on the basis of a three-segmented coelom but they have recently been shown to be multi-segmented.[38] The posterior region of the body, the opisthosoma, is morphologically similar to that of most annelids, apart from its lack of the gut. It is divided by muscular septa into coelomate compartments each bearing chaetae. These both appear early in ontogeny and the triple division of the coelom is considered to be only a transitory stage in development. In a study of the structure of the chaetae of pogonophores, the above authors have shown how their structure is similar to those of

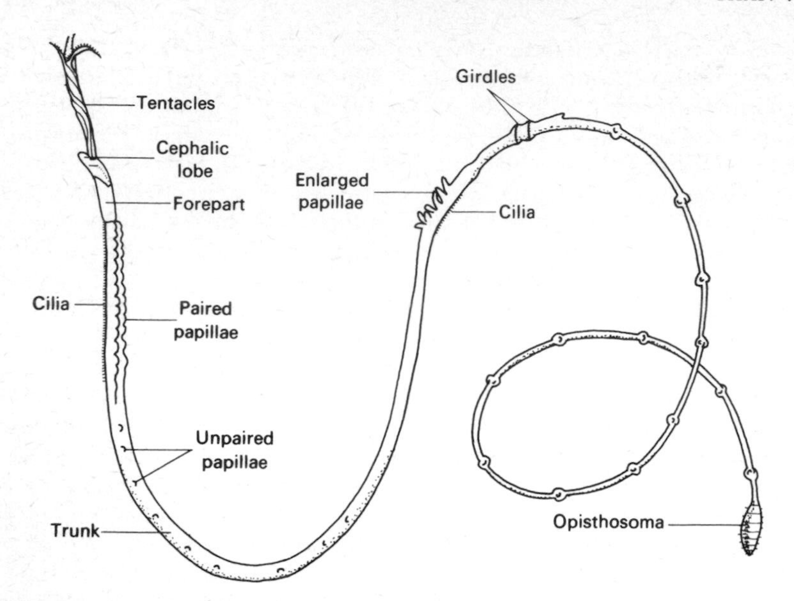

Fig. 7.7 Diagram of typical pogonophore, foreshortened, to show main body regions (from George and Southward[38]).

polychaetes and the evidence presented strongly supports the possibility of affinity with annelids. A typical adult pogonophoran has two rings of chaetae around the middle of the body and several rings around the opisthosoma (Fig. 7.7). The functions of the chaetae are to provide an anchor for movement in the burrow. The opisthosoma is used as a terminal burrowing organ, being alternately thrust through the substrate in a narrow extended form and dilated to anchor the tail with the aid of chaetal protrusion, so allowing the body to be pulled down into the substrate. George and Southward[38] point out how this process exactly resembles that of some sedentary polychaetes, for example a cirratulid, *Chaetozone.* Such similarity could, of course, be explained as separate but convergent evolution in a common habitat. However, the chaetae are very similar to those of annelids, although the specific design of an annelidan chaeta is not an essential tool for burrowing, for many animals burrow successfully without such a structure. George and Southward draw the conclusion that the chaetae are derived from a common 'proto-annelid' indicative of a close annelid-pogonophoran relationship. This and other evidence outlined by Southward,[102] would imply that the Pogonophora should be placed near the base of coelomate animals with close affinities to the Annelida (Fig. 7.6). However, the differences in morphology and development between the pogonophores, annelids, lophophorate phyla and primitive deuterostomes, such as the pterobranchs, suggests that these groups may have arisen separately from early coelomates and that their apparent similarities are indicative of a comparable grade of organization rather than of close affinity.

ORIGINS OF THE MAJOR PHYLA

Annelida

There can be little dispute of the gross morphology of the ancestral form of the annelids, for this group has clearly arisen from a primitive coelomate with the advent of metameric segmentation in response to the problems of burrowing in consolidated substrates. The characteristic annelidan form is that of a metamerically segmented animal exhibiting serial repetition of organs with functionally separate coelomic cavities in each segment, so affording localized changes in shape by the antagonism of longitudinal and circular muscles.

The Polychaeta have generally been considered the most primitive annelidan class largely on the basis of the primitive nature of their reproductive and excretory systems, particularly in comparison with the Oligochaeta. However, the arrangement of longitudinal and circular muscles of the body wall, and of the chaetae and septa, is much more primitive in the oligochaetes. These structures, moreover, represent the fundamental locomotory equipment of annelids. The development of a septate condition in an early coelomate requires only relatively minor changes in locomotory behaviour. Forward motion can only be achieved in a septate worm by retrograde waves and these were the most likely form of muscular locomotion in early Bilateria (p. 168). Much more fundamental changes to the basic locomotory pattern are required in respect of the polychaetes, with the development of parapodia and powerful pairs of dorsal and ventral longitudinal muscles (Fig. 5.1). The propellor surface of the worm has now become the parapodium rather than the body wall, and parapodial muscles have developed to effect a power stroke, as in *Nephtys* (Fig. 5.12), additional to the undulatory movements of the body caused by the longitudinal muscles. Locomotion with parapodia is well suited for movement through soft muddy substrates, a habit quite different from life in consolidated media, in which anchorage can be obtained by dilation of the entire body wall. This suggests a change of habitat between the earliest annelid and the development of parapodia. Evidence supporting this may be obtained from study of the evolution of the stomodaeum.[29]

The intersegmental septa probably initially had an important role in preventing high pressure in adjacent segments from affecting parapodia because of the thin nature of the parapodial wall; but with the development of parapodial muscles the septa are reduced or acquire different functions, such as serving to suspend the gut as in the Nereidae. Nevertheless, despite loss of septa, most errant polychaetes have retained a spacious coelom to maintain body turgor, probably because of the problems of obtaining stable muscle insertions without a hard skeleton. This requirement has been met by bracing points of the body wall with other muscles, the whole being kept taut during use by the coelomic hydraulic system. The evolution of a complex pattern of muscles in association with the parapodium, characteristic of the errant polychaetes, is in some ways comparable to the skeletal arrangements in arthropods.

Many changes in Annelida are regressive, however. Some polychaetes have reverted almost to the unsegmented coelomate condition, for example *Arenicola* or *Polyphysia*, while leeches have returned to a condition that is mechanically similar to acoelomate worms, characteristically creeping over surfaces in a looping motion employing anterior and posterior suckers for attachment. Looping movements involve the alternate contraction of all circular and all longitudinal muscles, thus avoiding localized muscle antagonism; thus the reason for segmentation is lost (p. 22), whilst the coelom is occluded by botryoidal tissue. Further details of the annelidan form and evolution are conveniently summarized by Dales.[30]

Arthropoda

This phylum is closely related to the Annelida and appears to have been derived from septate coelomate worms, since signs of segmentation are readily recognized in arthropods.[72] From a functional viewpoint the arthropods represent the culmination of the changes we have observed in annelids, for with the development of a rigid exoskeleton the hydrostatic function of body wall muscles and coelom is lost and replaced by muscle system characteristic of animals with hard skeletons. The coelom has now no locomotory function and is vestigial in the adult arthropod, only remaining as the gonadial cavity and in the paired excretory organs. Where the cuticle is soft, as in larvae or in the Onychophora, the haemocoel, rather than the coelom, acts as a fluid skeleton in locomotion in a manner comparable to that of other soft-bodied animals.

Embryological evidence as to the segmented organization of the coelom in arthropods indicates that the coelom was once much larger and has suffered reduction. This view was taken by Ray Lankester,[63] who suggested that the peripheral circulation expanded at the expense of the coelom, a process he called phlebodesis. This implies expansion of the haemocoel simultaneously with reduction of the coelom. If the coelom disappeared as a consequence of the secretion of a hard exoskeleton, we may ask why develop a haemocoel? Apart from in arthropods with soft cuticles, what can be the function of the haemocoel? Tiegs and Manton[108] hold the view that the Onychophora were the primitive stock from which myriapods and insects were evolved. *Peripatus*, an example of the Onychophora, has a soft cuticle and extensive haemocoel. If it is considered primitive to other arthropod groups then replacement of the coelom must have predated the hardening of the cuticle.

Manton[71] believes that the increase of the haemocoel occurred in a septate coelomate for otherwise why should arthropods be segmented? Certainly great exploitation of exoskeleton and haemocoel has led to many important steps of diversification resulting in the evolution of well defined arthropod groups. It is the initial steps of the radical change from coelom to haemocoel and the reasons for it that are obscure.

Some light is thrown on this question by Manton's[67] studies on *Peripatus* which crawls by movement of serial appendages without any longitudinal body muscles being directly involved. However, change in velocity results from a change of locomotory gait, which in turn is dependent on body length. Increased speed is associated with an extended body, and reduced speed with a shortened body. Such length changes involve the fluid skeleton and Manton argues that a segmented coelom only allows local changes in shape but that a single haemocoelic cavity operates uniformly throughout the body and is thus more suited. But this situation could be achieved just as easily by the breakdown of intersegmental septa, as in many annelids, and the retention of the coelom. Manton[69,70] has subsequently demonstrated further properties of the haemocoel in geophilomorph centipedes which may have some relevance to its origins. These animals, for example *Orya*, burrow into crevices in the soil in the manner of earthworms, by inserting the head in a crevice and enlarging it by radial thrust. Manton devised an experimental situation in which *Orya* lifted a platform of known weight; by knowing the area of the centipede in contact with the platform, she deduced the internal pressure. She obtained values of 3.6–4 N cm^{-2} for *Orya* and 1.3–2 N cm^{-2} in similar experiments using *Arenicola*. Seymour (1970) has repeated these experiments with *Arenicola* and has simultaneously measured internal pressure (Fig. 3.20). He has shown that the thrust on the platform is greatly in excess of that calculated directly from the pressure recorded from the body cavity and suggests that the difference is due to the stiffness of the body wall. On these grounds an internal pressure of about 2 N cm^{-2} might be expected in *Orya*. Manton considered it likely that high pressures (or maximum thrust) could only be attained once or twice in a coelomate because of leakage of fluid through the nephridia to the exterior at high pressure. However, this limitation would not apply to a haemocoel since it has no external openings. Moreover, while it is possibly true at the maximum level of pressure reported for *Orya*, we should remember that *Arenicola* repeatedly develops pressure pulses of the order of 1.5 N cm^{-2} during normal burrowing. Haemocoel and aseptate coelom function in the same manner in a hydraulic system, the pressure developed being dependent on the power of the muscles associated with the cavity. The supposed advantage of a haemocoel in not having direct communication to the exterior, and thus avoiding the possibility of leakage of body fluid, seems scarcely adequate to justify the change in the nature of the body cavity.

We may conclude that the arthropods evolved from an early pre-annelidan stock with segmented coelom by development of the cuticle, loss of the coelom and expansion of the haemocoel (Fig. 7.5). However, the functional reasons for this latter change cannot yet be satisfactorily explained.

Mollusca

The problem of the origin of the molluscs is similar in many ways to that of the arthropods for both possess an extensive haemocoel and an exoskeleton in the form of shell or cuticle respectively. One of the most important differences between these structures is their method of growth. In arthropods this occurs in stages by moulting whereas the molluscan shell grows more or less continuously by peripheral enlargement. Once secreted, the shell normally remains throughout life. This restricts flexibility in the molluscan form, except where the shell is reduced or absent and, indeed, it is this condition that we see in the most advanced examples of the phylum, such as the octopus.

The developmental stages of molluscs and polychaetes show great similarity in respect of both spiral cleavage and the form of the trochophore larva, suggesting close affinity between these groups. However, molluscs show neither the need of a coelomic cavity for locomotion nor signs of segmentation, with the exception of the monoplacophoran *Neopilina*.[64] Both of these features are, of course, characteristic of oligochaete and polychaete worms. Close association with the Annelida would thus indicate reduction of the coelom and expansion of the haemocoel but there are good reasons for the belief that expansion of the haemocoel is a primary one in molluscs in contrast to its secondary expansion in arthropods.[41] We must accordingly look elsewhere for molluscan origins.

Salvini-Plawen[95] has pointed out the similarity of many anatomical features of the Mollusca with turbellarian-like worms, while Graham[42] describes how the epithelium of molluscan skin is ciliated and bears a close resemblance to that of free-living flatworms and nemertines. Such an ancestor may be envisaged as being very thin dorso-ventrally, to move by cilia, to have its body spaces filled with parenchymatous tissue and to exhibit some serial repetition or pseudometamerism of gut diverticula and gonads along the length of its body. To evolve into a mollusc, development of gut and visceral mass would have occurred, resulting in dorso-ventral thickening and the specialization of the ventral tissues for locomotion. A flatworm must retain its shape because of the need for oxygen to diffuse to all tissues, so with the development of a visceral mass and the secretion of a calcareous shell, respiratory and circulatory organs became essential and all these structures probably evolved in parallel with each other. Initially the respiratory organs possibly consisted of numerous pairs of gills or ctenidia, as in *Neopilina* (Fig. 7.8), but these were reduced to a single pair in the stock from which the principal classes of Mollusca evolved.[37] The shell probably arose for purposes of protection first as a mucoid secretion, subsequently stabilized by quinone tanning to form an outer membrane, the periostracum, prior to calcification taking place to produce the calcareous shell characteristic of contemporary molluscs.[12,104]

With the secretion of the shell as a supporting and protective structure, what need was there for a coelomic cavity or segmentation? The only chamber in a mollusc that can be considered as coelomic is the pericardium but to do so requires that the gonocoel theory of coelomic origin is accepted at least in respect of the Mollusca. The pericardium is regarded as a dilation of the genital ducts and has undoubtedly arisen separately from the coelom of annelid worms. It clearly has no locomotory function and provides space for the beating of the heart.[37] The circulation of many molluscs is an open system in which the blood bathes organs in a large haemocoelic sinus and characteristically returns to the heart at very low or zero pressures. In consequence, there is little residual pressure for refilling the ventricle, which is brought about instead by the pericardium functioning as a constant volume system. Thus ventricular systole results in blood being forced out of the cavity through the aortae so tending to reduce the volume contained in the pericardium and generating a small negative pressure pulse. This negative pressure sucks oxygenated blood into the heart from the gills or mantle.[56] The pericardium thus plays a dynamic role in the circulation of the blood and should be considered as having evolved for this purpose rather than for locomotion.

It is only from studies of locomotion and the structure of the foot that we may be able to envisage the importance of the haemocoel and the possible needs of segmentation in the Mollusca. As in flatworms, locomotion is effected by cilia on the sole of the foot in most small molluscs, including juvenile forms, and those of small mass, such as the air-breathing freshwater pulmonates. With increase in size and weight, muscular pedal waves have taken over. Primitively, as in chitons and most prosobranchs, these are retrograde waves, just as in planarians and nemertines. It is only in the more specialized terrestrial snails and slugs that direct locomotory waves commonly occur.[58] Locomotion is much modified in the adults of bivalve molluscs for an infaunal mode of life but in larval stages it remains in a much more primitive state. Immediately after settlement of the free-swimming larvae onto a substrate the pediveliger moves by cilia, for example in *Venus*,[4] while soon after settlement of the epifaunal *Mytilus edulis* muscular locomotion, comparable to a retrograde pedal wave, has been observed by Lane (personal communication).

Except for the much modified cephalopods, cilia or pedal muscular locomotory waves are thus of fundamental importance to all molluscs as a means of movement over hard surfaces and the significance of a haemocoel consisting of numerous vesicles in association with pedal locomotory waves has already been emphasized (p. 30). In this phylum there is no locomotory requirement for a large fluid-filled cavity except in those forms which have become specialized for burrowing by expansion of the pedal haemocoel, as in many bivalves such as *Ensis*. Infaunal gastropods occur more uncommonly but either show an expansion of the haemocoel comparable to bivalves, as in *Bullia*,[20] or a large water filled cavity in the foot, as in *Natica* (p. 74).

The molluscan coelom only attains large dimensions in cephalopods, where pericardial and renopericardial cavities are almost coextensive with the mantle cavity in the posterior region of the body. Contraction of the circular mantle muscles during the jet cycle generates high pressure pulses in the mantle cavity and also compresses the viscera. This is comparable to the effect of adduction of the valves of bivalves where blood in the haemocoel is displaced ventrally to cause pedal dilation. A similar movement of blood in a cephalopod would tend to dilate the head, an effect incompatible with the high level of neural organization in this class. However, the extensive coelom around the viscera may function to localize the effect of a pressure pulse and reduce this effect.[123]

The view that molluscs were evolved from segmental animals would have received small support before the discovery of *Neopilina*, in which serial repetition of various organs occurs (Fig. 7.8).

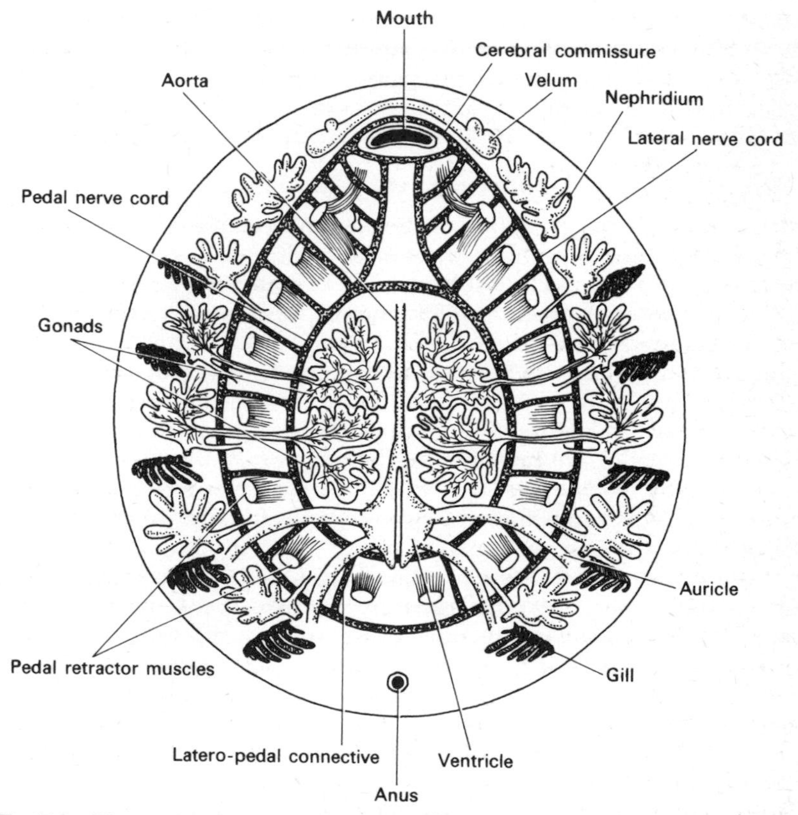

Fig. 7.8 Diagram of *Neopilina* (Monoplacophora) showing circulatory system and serial repetition of pedal retractor muscles, nerves, gills, reproductive and excretory organs (after Lemche and Wingstrand[64]).

However, it is uncertain whether this represents anything more than pseudometamerism comparable to that occurring in some turbellarians and nemertines. If segmentation were to be accepted as basic to Mollusca then we must look upon the ancestral mollusc as being a segmented coelomate in which the acquisition of a shell led to loss of segmentation and regressive evolution, particularly in respect of the foot. There could have been little mechanical significance in the segmented coelomate condition other than for locomotion and yet the primitive method of locomotion in molluscs is undoubtedly similar to free-living flatworms without any large fluid-filled body cavity. Some small pre-annelidan bilaterally symmetrical acoelomate worm, which exhibited serial repetition of some organ systems, seems a more likely origin for molluscs than does a coelomate segmented worm (Fig. 7.5). The presence of a trochophore larva in both annelids and molluscs indicates the comparability of the grade of organization of these two phyla and emphasizes their fundamental affinity at the base of the Protostomia.

Hemichordata, Echinodermata and Chordata

During the early evolution of the oligomerous deuterstomes three main phyla were established, the Hemichordata, Echinodermata and the Chordata (Fig. 7.6) and I will conclude with only brief comment on their evolution. The hemichordates are a small group of worm-like animals divisible into two closely related classes, the Enteropneusta and the Pterobranchia. Lophophore-like arms bearing tentacles occur upon the mesosome of pterobranchs but have been lost by enteropneusts on adoption of a burrowing habit. The presence of a lophophore makes the position of the Pterobranchia clear as the more primitive class of hemichordates, and some authors, for example Hyman,[54] also consider the pterobranchs to be ancestral to the echinoderms.

The derivation of the Chordata is difficult and to some extent speculative since the most primitive group, the Urochordata or tunicates, is almost completely sessile and considerably modified. However, the identical manner in which pharyngeal gill slits are formed and the assumption that gill slits only originated on one occasion suggests a close affinity between the hemi- and urochordates. Indeed, we can hardly avoid the conclusion that gill slits may have been first formed in animals resembling modern pterobranchs.[26] The three coelomic divisions, important in such an early deuterostome, have regressed, lost hydraulic functions and, apart from the metacoel have virtually disappeared in chordates.

The most important early developments in freely swimming chordates were nevertheless still concerned with locomotion despite the change of function of the coelom. It has been argued that the evolution of a neotenous tadpole-like larva in the tunicates has avoided the evolutionary *cul de sac* of a specialized sessile life characteristic of many urochordates.[14,16] The most significant developments in early chordates were the evolution of an internal skeletal system and segmental musculature. Yet even at this level of animal organization the importance of the fluid skeleton should not be overlooked, for we have already discussed (chapter 5) the manner in which the notochord of amphioxus functions as a fluid-muscle system.[45]

References

1. ALEXANDER, R. McN. (1966). Rubber-like properties of the inner hinge-ligament of the Pectinidae. *J. exp. Biol.*, **44**, 119.
2. ALEXANDER, R. McN. (1968). *Animal Mechanics*, Sidgwick & Jackson, London.
3. ALEXANDER, R. McN. (1971). *Size and Shape*, Arnold, London.
4. ANSELL, A. D. (1962). Observations on burrowing in the Veneridae (Eulamellibranchia). *Biol. Bull. mar. biol. Lab., Woods Hole*, **123**, 521.
5. ANSELL, A. D. (1967). Leaping and other movements in some cardiid bivalves. *Anim. Behav.*, **15**, 419.
6. ANSELL, A. D. (1970). Boring and burrowing mechanisms in *Petricola pholadiformis* Lamarck. *J. exp. mar. Biol. Ecol.*, **4**, 211.
7. ANSELL, A. D. and NAIR, N. B. (1966). A comparative study of bivalves which bore mainly by mechanical means. *Am. Zool.*, **9**, 857.
8. ANSELL, A. D. and TREVALLION, A. (1969). Behavioural adaptations of intertidal molluscs from a tropical beach. *J. exp. mar. Biol. Ecol.*, **4**, 9.
9. ANSELL, A. D. and TRUEMAN, E. R. (1968). The dynamics of burrowing of the sand dwelling anemone, *Peachia hastata*. *J. exp. mar. Biol. Ecol.*, **2**, 124.
10. ANSELL, A. D. and TRUEMAN, E. R. (1974). The energy cost of migration of the bivalve *Donax* on tropical sandy beaches. *Mar. Behav. Physiol.*, **2**, 21.
11. BATHAM, E. J. and PANTIN, C. F. A. (1950). Muscular and hydrostatic action in the sea anemone *Metridium senile* (L.). *J. exp. Biol.*, **27**, 264.
12. BEEDHAM, G. E. and TRUEMAN, E. R. (1967). The relationship of the mantle and shell of the Polyplacophora in comparison with that of other Mollusca. *J. Zool., Lond.*, **151**, 215.

13. BERRILL, N. J. (1950). *The Tunicata.* Ray Soc., London.
14. BERRILL, N. J. (1955). *The Origin of the Vertebrates.* Clarendon Press, Oxford.
15. BOARD, P. A. (1970). Some observations on the tunnelling of shipworms. *J. Zool., Lond.*, **161**, 193.
16. BONE, Q. (1960). The origin of the chordates. *J. Linn. Soc. (Zool.)*., **44**, 252.
17. BONE, Q. and RYAN, K. P. (1973). The structure and innervation of the locomotor muscles of salps (Tunicata: Thaliacea). *J. mar. biol. Ass. U.K.*, **53**, 873.
18. BOYCOTT, B. B. (1958). The cuttlefish—*Sepia. New Biology*, **25**, 98.
19. BRAND, A. R. (1972). The mechanism of blood circulation in *Anodonta anatina* (L.). (Bivalvia, Unionidae). *J. exp. Biol.*, **56**, 361.
20. BROWN, A. C. (1964). Blood volumes, blood distribution and sea water spaces in relation to expansion and retraction of the foot of *Bullia. J. exp. Biol.*, **41**, 837.
21. CARRICKER, M. R. (1961). Comparative functional morphology of boring mechanisms in gastropods. *Am. Zool.*, **1**, 263.
22. CHAPMAN, G. (1958). The hydrostatic skeleton in the invertebrates. *Biol. Rev.*, **33**, 338.
23. CHAPMAN, G. (1971). The movement and locomotion of invertebrates. In *The Invertebrate Panorama*, Smith, J. E. et al. Weidenfeld & Nicolson, London.
24. CHAPMAN, G. and NEWELL, G. E. (1947). The role of the body fluid in relation to movement in soft-bodied invertebrates. I. The burrowing of *Arenicola. Proc. roy. Soc.*, B. **134**, 431.
25. CLARK, R. B. (1963). Evolution of the coelom and metameric segmentation. In *The Lower Metazoa*, Ed. Dougherty, E. C., University of California Press, Berkeley.
26. CLARK, R. B. (1964). *The Dynamics of Metazoan Evolution.* Clarendon Press, Oxford.
27. CLARK, R. B. and CLARK, M. E. (1960). The ligamentary system and the segmental musculature of *Nephtys. Q. J. micr. Sci.*, **101**, 149.
28. CLARK, R. B. and COWEY, J. B. (1958). Factors controlling the change of shape of certain nemertean and turbellarian worms. *J. exp. Biol.*, **35**, 731.
29. DALES, R. P. (1962). The polychaete stomodaeum and the interrelationships of the families of Polychaeta. *Proc. zool. Soc. Lond.*, **139**, 389.
30. DALES, R. P. (1963). *Annelids*, Hutchinson, London.
31. DUVAL, D. M. (1963). The biology of *Petricola pholadiformis* Lamark (Lamellibranchiata, Petricolidae). *Proc. malac. Soc. Lond.*, **35**, 89.
32. EGGERS, F. (1935). Zur Bewegungs physiologie von *Malacobdella grossa* Müll. *Z. wiss. Zool.*, **147**, 101.
33. ELDER, H. Y. (1972). Connective tissues and body wall structure, of the polychaete *Polyphysia crassa* and their significance. *J. mar. biol. Ass. U.K.*, **52**, 747.

34. ELDER, H. Y. (1973a). Direct peristaltic progression and the functional significance of the dermal connective tissues during burrowing in the polychaete *Polyphysia crassa* (0ersted). *J. exp. Biol.*, **58**, 637.
35. ELDER, H. Y (1973b). Distribution and functions of elastic fibers in the invertebrates. *Biol. Bull. mar. biol. Lab. Woods Hole*, **144**, 43.
36. FLOOD, P. R. (1970). The connection between spinal cord and notochord in amphioxus (*Branchiostoma lanceolatum*). *Z. Zellforsch.*, **103**, 115.
37. FRETTER, V. and GRAHAM, A. (1962). *British Prosobranch Molluscs.* Ray Soc., London.
38. GEORGE, J. D. and SOUTHWARD, E. C. (1973). A comparative study of the setae of Pogonophora and polychaetous Annelida. *J. mar. biol. Ass. U.K.*, **53**, 403.
39. GLADFELTER, W. B. (1972). Structure and function of the locomotory system of *Polyorchis montereyensis* (Cnidaria, Hydrozoa). *Helgoländer wiss Meeresunters*, **23**, 38.
40. GOODBODY, I. and TRUEMAN, E. R. (1969). Observations of the hydraulics of *Ascidia. Nature, Lond.*, **224**, 85.
41. GRAHAM, A. (1955). Molluscan diets. *Proc. malac. Soc. Lond.*, **31**, 144.
42. GRAHAM, A. (1957). The molluscan skin with special reference to prosobranchs. *Proc. malac. Soc. Lond.*, **32**, 135.
43. GRAY, J. (1968). *Animal Locomotion*, Weidenfeld & Nicolson, London.
44. GRAY, J. and LISSMANN, H. W. (1964). The locomotion of namatodes. *J. exp. Biol.*, **41**, 135.
45. GUTHRIE, D. M. and BANKS, J. R. (1970). Observations on the function and physiological properties of a fast paramyosin muscle—the notochord of amphioxus (*Branchiostoma lanceolatum*). *J. exp. Biol.*, **52**, 125.
46. HADŽI, J. (1963). *The Evolution of the Metazoa.* Pergamon Press, Oxford.
47. HAMMOND, R. A. (1966). Changes of internal hydrostatic pressure and body shape in *Acanthocephalus ranae. J. exp. Biol.*, **45**, 197 and 203.
48. HAMMOND, R. A. (1970). The burrowing of *Priapulus caudatus. J. Zool., Lond.*, **162**, 469.
49. HAND, C. (1959). The origin and phylogeny of the coelenterates. *Syst. Zool.*, **8**, 191.
50. HARRIS, J. E. (1957) and CROFTON, H. D. (1957). Structure and function in the namatodes: internal pressure and cuticular pressure in *Ascaris. J. exp. Biol.*, **34**, 116.
51. HATSCHEK, B. (1888). *Lehrbuch der Zoologie.* Fischer, Jena.
52. HOGGARTH, K. R. and TRUEMAN, E. R. (1967). Techniques for recording the activity of aquatic invertebrates. *Nature, Lond.*, **213**, 1050.
53. HYMAN, L. H. (1951). *The Invertebrates II.* McGraw-Hill, New York.
54. HYMAN, L. H. (1955). *The Invertebrates V.* McGraw-Hill, New York.
55. JOHNSON, W., SODEN, P. D. and TRUEMAN, E. R. (1972). A study in jet propulsion: an analysis of the motion of the squid, *Loligo vulgaris. J. exp. Biol.*, **56**, 155.
56. JONES, H. D. (1971). Circulatory pressures in *Helix pomatia* L. *Comp. Biochem. Physiol.*, **39A**, 289.

57. JONES, H. D. (1973). The mechanism of locomotion of *Agriolimax reticulatus* (Mollusca: Gastropoda). *J. Zool., Lond.*, **171**, 489.
58. JONES, H. D. (1974). Locomotion. In *Pulmonate molluscs*, Ed. Fretter, V., Academic Press, London and New York.
59. JONES, H. D. and TRUEMAN, E. R. (1970). Locomotion of the limpet, *Patella vulgata* L. *J. exp. Biol.*, **52**, 201.
60. KNIGHT-JONES, E. W. (1952). On the nervous system of *Saccoglossus cambrensis* (Enteropneusta). *Phil. Trans.* B, **236**, 315.
61. LAGLER, K. R., BARDACH, J. E. and MILLER, R. R. (1962). *Ichthyology.* Wiley, New York.
62. LANE, F. W. (1957). *Kingdom of the Octopus.* Jarrolds, London.
63. LANKESTER, E. R. (1904). Structure and classification of the Arthropoda. *Quart. J. micr. Sci.*, **47**, 523.
64. LEMCHE, H. and WINGSTRAND, K. G (1959). The anatomy of *Neopilina galatheae*, Lemche, 1957. *Galathea Rep.*, **3**, 9.
65. LISSMANN, H. W. (1945). The mechanism of locomotion in gastropod molluscs. I. Kinematics. *J. exp. Biol.*, **21**, 58.
66. LISSMANN, H. W. (1946). The mechanism of locomotion in gastropod mulluscs. II. Kinetics. *J. exp. Biol.*, **22**, 37.
67. MANTON, S. M. (1950). The evolution of arthropodan locomotory mechanisms. Part I. The locmotion of *Peripatus. J. Linn. Soc. (Zool.)*, **41**, 529.
68. MANTON, S. M. (1961). Experimental zoology and problems of arthropod evolution. In *The Cell and the Organism*, Ed. Ramsay, J. A. and Wigglesworth, V. B., University Press, Cambridge.
69. MANTON, S. M. (1952). The evolution of arthropodan locomotory mechanisms. Part 2. General introduction to the locomotory mechanisms of the Arthropoda. *J. Linn. Soc. (Zool.)*, **42**, 93.
70. MANTON, S. M. (1965). The evolution of arthropodan locomotory mechanisms. Part 8. Functional requirements and body design in Chilopoda. *J. Linn. Soc. (Zool.)*, **45**, 251.
71. MANTON, S. M. (1970). Arthropods: Introduction. In *Chemical Zoology V*, Ed. Florkin, M. and Scheer, B. T. Academic Press, New York.
72. MANTON, S. M. (1973). Arthropod phylogeny—a modern synthesis. *J. Zool., Lond.*, **171**, 111.
73. MARTIN, R. (1966). On the swimming behaviour and biology of *Notarchus punctatus* Phillipi (Gastropoda, Opisthobranchia). *Pubbl. staz. zool. Napoli*, **35**, 61.
74. METTAM, C. (1967). Segmental musculature and parapodial movement of *Nereis diversicolor* and *Nephtys hombergi* (Annelida: Polychaeta). *J. Zool., Lond.*, **153**, 245.
75. METTAM, C. (1971). Functional design and evolution of the polychaete *Aphrodite aculeata. J. Zool., Lond.*, **163**, 489.
76. MILL, P. J. and PICKARD, R. S. (1972). Anal valve movement and normal ventilation in aeshnid dragonfly larvae. *Odonatologica*, **1**, 41.
77. MOORE, J. D. and TRUEMAN, E. R. (1971). Swimming of the scallop, *Chlamys opercularis* (L.). *J. exp. mar. Biol. Ecol.*, **6**, 179.

78. MORTON, J. E. (1954). The biology of *Limacina retroversa*. *J. mar. biol. Ass. U.K.* **33**, 297.
79. MOSELEY, H. N. (1877). Notes on the structures of several forms of land planarians, with a description of two new genera and several new species, and a list of all species at present known. *Quart. J. micr. Sci.*, **17**, 273.
80. NAIR, N. B. and ANSELL, A. D. (1968a). Characteristics of penetration of the substratum by some marine bivalve molluscs. *Proc. malac. Soc. Lond.*, **38**, 179.
81. NAIR, N. B. and ANSELL, A. D. (1968b). The mechanism of boring in *Zirphaea crispata* (L.) (Bivalvia: Pholadidae). *Proc. roy. Soc.*, B, **170**, 155.
82. NEWELL, G. E. (1950). The role of the coelomic fluid in the movements of earthworms. *J. exp. Biol.*, **27**, 110.
83. OLIVO, R. F. (1970). Mechanoreceptor function in the razor clam: sensory aspects of the foot withdrawal reflex. *Comp. Biochem. Physiol.*, **35**, 761.
84. OWEN, G. (1966). Feeding. In *Physiology of Mollusca*, I, Ed. Wilbur, K. M. and Yonge, C. M. Academic Press, London and New York.
85. PACKARD, A. (1966). Operational convergence between cephalopods and fish: an exercise in functional anatomy. *Arch. Zool. Ital.*, **51**, 523.
86. PACKARD, A. (1969). Jet propulsion and the giant fibre response of *Loligo*. *Nature, Lond.*, **221**, 875.
87. PACKARD, A. (1972). Cephalopods and fish: the limits of convergence. *Biol. Rev.*, **47**, 241.
88. PACKARD, A. and TRUEMAN, E. R. (1974). Muscular activity of the mantle of *Sepia* and *Loligo* (Cephalopoda) during respiration and jetting and its physiological interpretation, *J. exp. Biol.*, **61**, 411.
89. PANTIN, C. F. A. (1950). Locomotion of nemertines. *Proc. Linn. Soc. Lond. (Zool.)*, **162**, 23.
90. PANTIN, C. F. A. (1960). Diploblastic animals. *Proc. Linn. Soc. Lond. (Zool.)*, **171**, 1.
91. PEARL, R. (1903). The movements and reactions of freshwater planarians: a study in animal behaviour. *Quart. J. micr. Sci.*, **46**, 509.
92. PURCHON, R. D. (1955). The structure and function of the British Pholadidae (rock-boring Lamellibranchia). *Proc. Zool. Soc. Lond.*, **124**, 859.
93. RUNHAM, N. W. and HUNTER, P. J. (1970). *Terrestrial Slugs*. Hutchinson, London.
94. RUSSELL-HUNTER, W. D. and RUSSELL-HUNTER, M. (1968). Pedal expansion in the naticid snails. I. Introduction and weighing experiments. *Biol. Bull. mar. Lab., Woods Hole*, **135**, 548.
95. SALVINI-PLAWEN, L. VON (1968). Die 'functions-coelomtheorie' in der evolution der Mollusken. *Syst. Zool.*, **17**, 192
96. SCHIEMENZ, P. (1884). Uber die Wasseraufnahme bei Lamellibranchiaten und Gastropoden. *Mitt. Zool. Stn. Neapel*, **5**, 509.

97. SCHOPF, J. W., HAUGH, B. N., MOLNAR, R. E. and SATTERTHWAITE, D. F. (1973). On the development of metaphytes and metazoans. *J. Paleontol.*, **47**, 1.
98. SEYMOUR, M. K. (1969). Locomotion and coelomic pressure in *Lumbricus terrestris* L. *J. exp. Biol.*, **51**, 47.
99. SEYMOUR, M. K. (1970). Skeletons of *Lumbricus terrestris* L. and *Arenicola marina* (L.), *Nature, Lond.*, **228**, 383.
100. SEYMOUR, M. K. (1971). Burrowing behaviour in the European lugworm *Arenicola marina* (Polychaeta: Arenicolidae). *J. Zool. Lond.*, **164**, 93.
101. SEYMOUR, M. K. (1973). Motion and the skeleton in small nematodes. *Nematologica*, **19**, 43.
102. SOUTHWARD, E. C. (1971). Recent researches on the Pogonophora. *Oceanogr. mar. Biol. Ann. Rev.*, **9**, 193
103. STANLEY, S. M. (1973). An ecological theory for the sudden origin of multi-cellular life in the late precambrian. *Proc. Nat. Acad. Sci. U.S.A.*, **70**, 1486.
104. STASEK, C. R. (1973). The molluscan framework. In *Chemical Zoology*, VIII, Ed. Florkin, M. and Scheer, B. T. Academic Press, New York.
105. SWEDMARK, B. (1964). The interstitial fauna of marine sand. *Biol. Rev.*, **39**, 1.
106. TAYLOR, G. (1952). Analysis of the swimming of long and narrow animals. *Proc. roy. Soc.*, A, **214**, 158.
107. THOMPSON, T. E. and SLINN, D. J. (1959). On the biology of the opisthobranch *Pleurobranchus membranaceus*. *J. mar. biol. Ass. U.K.*, **38**, 507.
108. TIEGS, O. W. and MANTON, S. M. (1958). The evolution of the Arthropoda. *Biol. Rev.*, **33**, 255.
109. TRUEMAN, E. R. (1953). Observations on certain mechanical properties of the ligament of *Pecten*. *J. exp. Biol.*, **30**, 453.
110. TRUEMAN, E. R. (1966a). Observations on the burrowing of *Arenicola marina* (L.). *J. exp. Biol.*, **44**, 93.
111. TRUEMAN, E. R. (1966b). The mechanism of burrowing in the polychaete worm, *Arenicola marina* (L.). *Biol. Bull. mar. biol. Lab. Woods Hole*, **131**, 369.
112. TRUEMAN, E. R. (1966c). The fluid dynamics of the bivalve molluscs *Mya* and *Margaritifera*. *J. exp. Biol.*, **45**, 369.
113. TRUEMAN, E. R. (1967). The dynamics of burrowing in *Ensis* (Bivalvia). *Proc. roy. Soc.*, B. **166**, 459.
114. TRUEMAN, E. R. (1968a). The locomotion of the freshwater clam *Margaritifera margaritifera* (Unionacea: Margaritanidae). *Malacologia*, **6**, 401.
115. TRUEMAN, E. R. (1968b). The mechanism of burrowing of some naticid gastropods in comparison with that of other molluscs. *J. exp. Biol.*, **48**, 663.
116. TRUEMAN, E. R. (1968c). The burrowing activities of bivalves. *Symp. zool. Soc. Lond.*, **22**, 167.

117. TRUEMAN, E. R. (1968d). Burrowing habit and the early evolution of body cavities. *Nature, Lond.*, **218**, 96.
118. TRUEMAN, E. R. (1969). The fluid dynamics of molluscan locomotion. *Malacologia*, **9**, 243.
119. TRUEMAN, E. R. (1970). The mechanism of burrowing of the mole crab, *Emerita. J. exp. Biol.*, **53**, 701.
120. TRUEMAN, E. R. (1971). The control of burrowing and the migratory behaviour of *Donax denticulatus* (Bivalvia: Tellinacea). *J. Zool., Lond.*, **165**, 453.
121. TRUEMAN, E. R. and ANSELL, A. D. (1969). The mechanism of burrowing into soft substrates by marine animals. *Oceanogr. Mar. Biol. Ann. Rev.*, **7**, 315.
122. TRUEMAN, E. R., BRAND, A. R. and DAVIS, P. (1966). The dynamics of burrowing of some common littoral bivalves. *J. exp. Biol.*, **44**, 469.
123. TRUEMAN, E. R. and PACKARD, A. (1968). Motor performances of some cephalopods. *J. exp. Biol.*, **49**, 495.
124. WALLACE, H. R. (1958). Movement of eelworms. I. *Ann. Appl. Biol.*, **46**, 74.
125. WALLACE, H. R. (1959). Movement of eelworms in water films. *Ann. Appl. Biol.*, **47**, 366.
126. WARD, D. V. (1972). Locomotory function of squid mantle. *J. Zool. Lond.*, **167**, 487.
127. WARD, D. F. and WAINWRIGHT, S. A. (1972). Locomotory aspects of squid mantle structure. *J. Zool., Lond.*, **167**, 437.
128. WEBB, J. E. (1973). Swimming in amphioxus. *J. Zool., Lond.*, **170**, 325.
129. WEBB, J. E. and HILL, H. B. (1958). The ecology of Lagos lagoon. IV. On the reactions of *Branchiostoma nigeriense* Webb to its environment. *Phil. Trans.* B, **241**, 355.
130. WELLS, G. P. (1954). The mechanism of proboscis movement in *Arenicola. Q. J. micr. Sci.*, **95**, 251.
131. WELLS, G. P. (1961). How lugworms move. In *The Cell and the Organism*, Ed. Ramsay, J. A. and Wigglesworth, V. B. University Press, Cambridge.
132. WELLS, M. J. (1962). *Brain and Behaviour in Cephalopods.* Heinemann, London.
133. YONGE, C. M. (1936). The evolution of the swimming habit in the Lamellibranchia. *Mem. Mus. Hist. nat. Belg.*, **3**, 77.
134. YONGE, C. M. (1955). Adaptation to rock boring in *Botula* and *Lithophaga* (Lamellibranchia, Mytilidae) with a discussion on the evolution of this habit. *Q. J. micr. Sci.*, **96**, 383.
135. YONGE, C. M. (1963). Rock-boring organisms. In *Mechanisms of Hard Tissue Destruction*, Ed. Sognnaes, R. F., Am. Ass. Adv. Sci., Washington.

Index

Major entries (page numbers) are shown in **bold** type

abrasive movement, 88, 100, 102, 104
Acanthocephala, 173
acceleration, 2, 4, 110, 129, **139–41**, 157
accessory boring organ, 89
acellular animals, 163
aciculum, 108–10, 113
Acoela, 18, 164, 167
acoeloid form, 164, 166
acoelomate worm, 162, 163, 166, 167, 170, 180, 186
Actinaria, 52
actinotroph larva, 177
adductor muscle, **63–9**, 76–7, 90, 93–7, 100–2, 104, 105
Adesmacea, 89, 90, 95, 104
adhesion, 39, 41, 56
aggregation, 163, 166
Agriolimax, 35–7
algae, 161
Amoeba, 24
amphioxus, 44, 70, **125–8**, 187
anchor, 19, 26, 27, 40, 41, 110, 178
anchorage, 21, **25–7**, **41–7**, 52, 53, 56, 60, 62, 64–9, 71–4, 160
anemone, 10, 13, 42, 44, 52–6, 76, 78, 150, 160
Annelida, 13, 43, 66, 76, 77–80, 117, 128, 162, 167, 170–2, 174–80, 183, 186
Anodonta, 66, 77
antagonism, **4–15**, 76, 93, 94, 143, 144, 148, 175, 180
antagonistic muscles, 11, 16, 34–6, 93, 143, 148
Anthozoa, 164, 171
anti-sea gull reflex of lugworm, 58
Aphrodite, 113–14
Aplysia, 116
apodous holothurians, 72, 73
archenteron, 170
Arenicola, 5, 7–10, 25, 27, 42–4, **48–53, 55–63**, 70–2, 74–80, 108, 160, 169, 175, 180, 182
Arthropoda, 1, 74, 112, 114, 162, 170, 176, **180–2**
Ascaris, 16, **117–19**, 173
Aschelminthes, 162, 164, 166, 173
Ascidia, 14, 15, 158
aseptate worm, 8, **44–7**, 60, 76, 163
Aspidosiphon, 87
autotroph, 161
axial skeleton, 107, **125–8**, 175

barnacle, 87
beet eelworm, 119
bilateral symmetry, 162, 167–9
Bilateria, 167–73
Bipalium, 22
bivalve molluscs, 15, 47, 48, 55, **63–71**, 73–7, **81–106**, 152, 154, 184
blood, 32–4, 66, 67, 75, 76, 97, 172, 183–4
blood vascular system, 172, 183–4
blastaea stage, 163
blastocoel, 173, 174
blastopore, 176, 177
blastula, 162, 163
body cavities, 43, 75, 76, 78, 160, 169, 170, 182
body wall, stiffness, 61, **79, 80**
boring, 87–106
botryoidal tissue, 22, 180
Botula, 90, 105
Brachiopoda, 176, 177
branchial chamber, 14, 15, 158
Branchiostoma, 125–8, 187
brittle star, 74
buccal mass, 49–51
Bullia, 73, 184
buoyancy, 157
burrow, 16, 27, 55, 60
burrow packing force, 61, 63, 72
burrowing, **40–87**, 100, 112, 127, 154, 169, **172–5**, 177, 178
burrowing by, Annelida, 44–52, 56–63, 78–80, 112
  Arthropoda, 43, 74
  Coelenterata, 52–6, 74
  Echinodermata, 43, 72, 73
  Mollusca, 40, 47, 48, 63–71, 73–7, 80–6, 93, 94
  Priapulida, 54, 55, 71, 72
  Sipunculida, 54, 55, 72
burrowing habit, 72–5, 167, 186

calcareous secretion, 87, 106
calcareous rock, 88, 105, 106
Cambrian period, 160–2
capitulum, 56
Cardiacea, 89
Cardium, 66
cartilagenous sockets, 132
Cassidae, 89
Cassiopeia, 157
cellularization, 164, 166
cellulolytic bacteria, 104
centipede, 181, 182
cephalic lobe, 177, 178
Cephalocordata, 171
Cephalopoda, **129–50**, 168, 184, 185
Cerebratulus, 72, 115, 169, 170
Cestoda, 164, 166, 176
chaeta, 5, 50–2, 88, **108–10**, 177, 178
chaetal extrusion, 52, 58, 61, 178
Chaetozone, 178
chalk, 90, 91
chemical solution, 88, 89, 105
chitons, 28, 168, 184
Chlamys, 152, 153
Chordata, 44, 107, **125–8**, 162, 171, 175, 176, 186, 187
cilia, **16–18**, 28, 100, 172, 178, 183, 184
ciliary movement, 16–18, 42
Ciliata, 19, 164
circular muscles, 5–7, **10–14**, 21, 22, 24, 25, **43–6**, 50–2, 56, 60–3, 72, 78, 94, 100, 108, 113, 117, 132, **143–50**, 157, 169, 179, 184
circulatory system, 18, 76, 172, **183–5**
clam, 43, 48, 75, 80–6
Cliona, 88, 104
Clione, 117
Cnidaria, 163–5
cockle, 74
coefficient of discharge, 133, 136, 137, 140–2, 157
Coelenterata, 10–14, 43, 75, 162, 164, 166, 167
coelenteric fluid, 12–14, 56
coelenteron, 11–14, 43, 56, 75, 160, 163, 166
coelom, 5, 8, 43, **49–52**, 71, 73, 75, **78–80**, 110, 132, 160, **176–6**, **179–87**
coelomate, 8, 25, 162, 167, 169, 170–73, 176, 179
coelomoduct, 7
coelomic fluid, 7, 46, 50–2, 56, 76, 78, 110, 132, 148
coelomic pressure, 50–2, 55, 57, 61
collagen, 125, 148
collagen fibres, 17, 142
collagen lattice, 17, 45, 142, 143
column of anemone, 12, 52, 53
contraction, 144, 148, 156
control of locomotion, 80–2, 148–50, 168, 172
constant volume system, **10–15**, 182, 184
Convoluta, 164
coral reef, 87, 106
crab, 43, 74, 150
cranchid squid, 144
crawling, 22, **27–41**, 63, **107–14**, 167, 183, 184
crevice burrowing, 60, **80**, 181
Crustacea, 160
Ctenophora, 163–6
cuticle, 15, 61, 117, 161, 180–2
cuttlefish, 130–3, 135–7, 143–9
cyphonautes larva, 177

Dendrocoelum, 19, 20
density, 19, 115, 133, 140
Dentalium, 55, 73
detritus, 167, 169, 172
Deuterostomia, 162, 171, 176–9, 186
digestive system, 165, 167, 172
digging cycle, **47–8**, 54, 57, **64–7**, 70–3, 75, 81, 82
digging period, **47–8**, 53, 55, 58, 70, 81, 82
dilation, 8, 26, 27, **46**, **47**, 55, 57, 60, **64–7**, 72, 91, 97, 160, 179, 185
dimyarian bivalve, 63
diploblastic animals, 163
dipleurula larva, 176
direct wave, 25, **27**, **35–41**, **44–6**, 72, 168, 184
discharge coefficient, 133, 136, 137, 140–2, 157
ditaxic locomotion, 28, 112
Diurodrilus, 167
Donax, 47, 48, 66–8, 75, 77, **80–6**, 127
dorso-ventral muscles, 17–19, 29–35, 41, 168
double fluid muscle system, 73, **76–7**, 129
double innervation of cephalopod mantle, 149, 150
drag, 75, 121, 132, 133, **140-2**, 154, 157
drag coefficient of squid, 140, 141
dragonfly larva, 129

earthworm, 8, 44, 46–7, **60–3**, 79, 80, 107, 112, 169, 174, 181
Echinodermata, 87, 162, 170–2, 176, 177, 186
Echinoidea, 87, 160
ectoderm, 165
Ectoprocta, 176, 177
eel, 114, 119, 120
effector mechanisms, 16
elastic fibres, 45, 46, 155
elastic ligament, 15, **63–6**, 73
electrodes, 14, 15, 144, 145
electronic recording techniques, 3, 63, 134, 144, 145
Eledone, 136, 140
Emerita, 48, 74, 75
endocrine system, 172, 173
endoderm, 165
Endoprocta, 173
energy requirement for,
burrowing, 75, 86
migration, 86
energy store, hinge ligament, 15, **63–6**, 73, **154**, **155**, 158
mesoglea, 13, 155–8
test, 14, 15
Ensis, 42, 43, **66–70**, **76–8**, 101, 160, 169, 184
enterocoelous development of coelom, 163, **170–2**, 176, 177
Enteropneusta, 70, 73, 171, 186
epipodia, 117
errant polychaetes, 72, 107–112, 117, 122–4
escape movements, *Arenicola*, 58
Cardium, 74
Chlamys, 129, 152–4, 157
Loligo, 142, 157

escape movements—*continued*
Notarchus, 129, 150, 151
Planaria, 20, 21
eversion of physa, proboscis, **49–51**, 52, 53, 55, 58
eversion–introversion cycle, 52, 53
excretory system, 173–5, 179, 180
exhalant opening/siphon, 90, 93–5, 130, 131, 158
exhalation, 132, 144
exoskeleton, 1, 31, 43, 102, 180–2
extension, **25–7**, 31, 35, 36, 40, 44, 46, 67, 68, 75, 168, 175
extensor muscle, 1
external hydraulic system, 34
extracellular digestion, 13
extrinsic muscles, 110

feeding tentacles, 174, 176–8, 186
fins, 136
fish, 114, 115, 117, 122, 129
fixation, 36
flagellates, 162, 163
flanges of *Arenicola*, 49–52, **58–60**
flange—proboscis sequence, 58, 59
flatworms, **16–25**, 41, 87, 167–9, 183, 184, 186
flexor muscle, 1, 110
fluid dynamics of burrowing, **75–80**
fluid-muscle system, **169–74**, 181, 187
in amphioxus, 125–8
in *Arenicola*, 7, 8, 58–60, 74, 76, 78–80
in *Ascaris*, 117, 128
in *Ensis*, 68, 69, 76–8
in *Loligo*, 135–9, 142, 148
in *Metridium*, 10–15
in *Natica*, 73, 74
in *Priapulus*, 7, 54, 55, 71
in salps, 158
in *Sepia*, 135–9, 143, 148
in *Sipunculus*, 7, 54, 55
foot of, bivalve molluscs, **63–70,** 76, 77, 81–5, 91, 94, 184
Ensis, 66, 67, 69, 70, 76, 77, 80
Patella, 29–35, 184
foot of,—*continued*
Polinices, 40, 41, 74
slug, 35–9, 184
snail, 28, 89, 184
Teredo, 102, 103
Zirphaea, 95–7
force, 2, 8, 19, **38–9**, 44, 48, 50, 51, 60, 61, 63, 69, 75–80, 110, 114, 115, 119, 127
force transducer, 3, 39, 70, 75, 80, 134
fossil forms, 160
funnel, 3, 130–3, 136–8
funnel, outlet area, 133, 136, 149

gait, 181
galloping movement of snails, 40, 41
gastraea stage, 163
theory of origin of Metazoa, 162, 163, 165
Gastrochaenacea, 89
Gastropoda, 27, 40, 41, 73, 88, 89, 104, 116, 117
Gastrotricha, 173
gastrula, 161, 163, 165
geodesic network, 17, 66, 117, 142, 143
Geoplana, 22, 23, 25
Geonemertes, 18
giant neurones, 148, 149
gill, 76, 77, 116, 129, 131, 172, 183, 185
gill slits, 186
Glycymeris, 66
gonads, 171–3, 183, 185
gonocoel theory, 171, 172
gravity, 133, 140
growth, 139, 140, 157, 167, 172, 173
gut, 13, 76, 118, 177, 183
gut movement in nematodes, 118

hard substrates, **87–106**
haemocoel, **28–36**, 41, 43, 56, **63–7**, 76, 77, 100, 104, 129, 160, 168, 180, 184
haemocoelic spaces, 144
Haemonchus, 118, 122
hard-bodied animals, 1, 74, 160, 180, 181
head, 25, 127
heart, 183, 185
heart urchin, 43
Helix, 24, **27–8**, 32, 40, 110–1
Hemichordata, 162, 176–7, 186
Heterodera, 119
heterotrophs, 161
Hiatella 90, 104
hinge, of scallops, 154, 155, 158
hinge teeth, 81
Hirudo, 21
Holothuria, 72, 167, 175
hormones, 172, 173
hooked squid, 142
Hydra, 163
hydraulic system, 1, 4, 8, **10–14**, 16, 52, **75, 76, 78**, 101–4, 160, 170–3, 180, 182
hydrostatic skeleton, 1, 5, 6, **10–16**, 22, 72, 75, 107, 125, 128, 158, 167, **169–72**, 174, 180
Hydrozoa, 155, 164, 167
hyperventilation, 143, **146–50**

inertial forces, 18, 19, 115, 132
inhalant opening/siphon, 90, 93, 94, 100, 130–2, 158
inhalation, 142–8, 158
initial penetration, 48–55, 160
Insecta, 181
internal hydraulic system, 34
interstitial fluid, 169, 170
intracellular digestion, 165, 168
intrinsic muscles, 110, 113
introversion of physa, 52, 53, 55, 56
isotonic transducer, 147

jellyfish, 130, **155–8**
jet cycle, 130 **134–7**, 139–42, 144–7, 149, 150, 154
jet orifices, 130–2, 150, 151, 153
jet propulsion, 3, **129–158**
jet swimming, 3, 129, 132 156, 157
jet thrust, **132–7**, 140–2, 150, 156, 157
jet velocity, 132, 133, 153

Keber's valve, 66, 76, 77
kidney, 77
Kinorhyncha, 173

laminar flow, 122
lancelet, 125–8
leech, **21–2**, 41, 71, 115, 122, 180
legs, 74, 107, 112
Leptoplana, 19, 115
Leptosynapta, 72, 73, 167, 175
lifting force, 79, 80
ligament, **63–6**, 73, 83, 90–5, 104, 105, **154**, 158
Limacina, 116, 117
Limax, 36
limbs, 74, 113, 114, 160
limpet, 28–35
Lineus, 18, 24, 25
Lithobius, 115
Lithophaga, 105
Lithotrya, 87
locomotion into substrates, 42–106
locomotion over substrates, **16–41**, **106–14**, 118, 119
locomotory waves, 20, **22–41**, 61
Loligo, 4, 130–7, **139–45**, 147, 149, 157
longitudinal muscle, 4–8, **10–15**, 21–4, 30, 34–8, 43–6, 50, 51, 60, 61, 72, 74, 107–13, **123–5**, 143–50, 168, 169, 175, 179, 180, 181
looping movements, 21–2, 41, 71, 150, 168
lophophorates, 170, 171, 174, 176, 177
lophophore, 176, 186
loping gait of snails, 40, 41
lugworm, 10, 25, 42, 48–53, 57, 60, 78
Lumbricus, 46, 60–3, 79, 80

Mactra, 66–8, 70
Malacobdella, 21
manometer, 9, 13, 52
mantle, 67, 77, 102, 151–4, 168
mantle cavity, 3, **64–9**, 103, 104, 129–40, **152–4**, 157, 184
mantle cavity capacity, 67, 132–3, 140, 157
mantle cavity pressure in, *Chlamys*, 152, 153
Ensis, 68, 76
Loligo, 130–42, 145, 157
Margaritifera, 68, 77
Sepia, 135, 136, 146
Octopus, 136–8
mantle cavity pressure in—*continued*
Teredo, 104
Zirphaea, 100, 101
mantle, lobes, 77, 78
mantle muscles, 3, 16, 77, 78, 130–3, **142–50**, 157, 185
mantle muscles, innervation in squid, 148–50
Margaritifera, 68, 77
mass, 2, 18
mechanical abrasion, 88–100, 104, 105
meiofauna, 167
Mercenaria, 69, 70, 104, 105
mesentery, 11–13
mesocoel, 174, 177
mesoderm, 16, 163, 170, 174
mesoglea, 11–13, 155–8
metacoel, 177
metameric segmentation, 8, 117, **124–8**, **174–6**, 178
Metaphyta, 161
Metazoa, 16, **159–76**
metazoan evolution, 128, **159–87**
Metridium, 10–15, 169
migratory behaviour of *Donax*, 83–6
migratory cycle, 84, 85
model, squid motion, 140–2
mole crab, 48, 74, 75
Mollusca, 25–43, 48, **63–70**, 73–8, 87, 89, 102, 129, 160, 162, **168–72**, 176, **182–6**
momentum, 3, 4, 132
monotaxic locomotion, 28, 112
mouth, 165, 167, 177, 185
mucous adhesion, 21, 36, 38, 41
mucous glands, 165, 168, 172
mucus, **18–23**, 31, **36–8**, 60, 88, 100, 183
mud, 45, 88, 158, 169, 179
Murex, 88
muscle tension, 10, 76–8, 138
muscle action potentials, 149
muscular antagonism, 1, 142–8
mussel, 184
Mya, 66–70, **93–4**, 101, 104–5
Myacea, 89, 90
myocommata, 125, 126
myopodia, 23
myotome, 25, 26
Myriapoda, 181
Mytilacea, 89
Mytilidae, 90

Nassarius, 28
Natica, 40, 55, 70, 73, 74, 88, 89, 104, 184
negative buoyancy, 157
negative pressure, 61, 184
Nematoda, 15, **117–19**, 122, 173
Nemertea, 7, **16–25**, 28, 41, 72, 76, 115, 117, 142, 162, 166, 168–70, 183–4, 186
nephridium, 7, 171, 182, 185
nephrocoel, 171
Nephtys, 55, 70, 72, 112, 113, **122–4**, 179
Neopilina, 183, 185, 186
neoteny, 163, 173, 187
Nereis, **107–13**, **122–4**, 180
nervous system, 165, 168, 172, 175, 185
nervous control of burrowing, 80–5
neurone, 149
neuropodium, 108, 109, 113
Newton's laws of motion, 2, 19, 132
neutral buoyancy, 144
nitrogen, liquid, 36
Notarchus, 129, 150–2
notochord, **125–8**, 175, 187
notopodium, 108–9, 113, 114
Nucula, 66

Octopus, 130, 133, **136–8**, 182
oblique muscle, 5, 21, 35–8, 108, 109
Oligochaeta, 179, 183
oligomerous animals, 170, **174–7**, 186
Onchoteuthis, 142
ontogeny, 157
Onychophora, 180, 181
Opisthobranchia, 129, 150
opisthosoma, 177, 178
oral sphincter muscle, 56
Ordovician period, 162
origin of Metazoa, 161–7
Orya, 181, 182
oyster shell, 88, 89

papillae, 49, 178
paramyosin, 125, 126
parapodial canal, 50–2
parapodial muscles, 108–10

parapodium, **107–14**, 122–4, 150, 151, 179, 180
parenchyma, **16–25**, 168, 169, 183
Patella, 28–35
Peachia, **42–4**, **52**, **53**, 56, 75, 160
Pectinacea, 152–4
pectoral fins, 129
pedal crawling in snails, 28–41
pedal epithelium, 30–6
pedal ganglion, 81
pedal haemocoel, 30–6, 67, 68, 71, 76
pedal locomotory waves, 20–41, 184
pedal musculature, 29–31, 35–41, 74, 76, 81
pedal protraction, 63–7, 76, 81, 82, 97
pedal retraction, 63–9, 76, 185
penetration anchor, **43–4**, 48, 54–6, 60, **64–7**, **70–4**, 91, 160, 175
pericardium, 132, 183, 184
Peripatus, 181
peristalsis, 22, **24–7**, 52, 53, 110–12, 169, 172, 177
peristaltic wave, **24–7**, 52, 53, 56, 72
periostracum, 89, 183
perivisceral coelom, 73
Petricola, 90–2, 104
pharynx, 49, 60, 118
phlebodesis, 181
Pholadidae, 95, 105
Phoronida, 176, 177
Phoronis, 177
photographic analysis, 32
phylogeny, 159–187
physa, 52, 53, 55, 56, 74
phytoplankton, 161
pike, 129
Placophora, 27
Planaria, **18–20**, 167–9, 184
plankton, 88, 161, 168
planula larva, 165, 166
planuloid–acoel theory of origin of Metazoa, 165, 166
Platyhelminthes, 16, 166–8, 170
Platyodon, 90, 94, 95
Pleurobranchus, 116, 117
Pogonophora, 171, 177–9
point d'appui, **23**, **24**, 26, 27, 39, 44, 62, 110, 168
Polinices, 40, 41, 73, 74
Polydora, 88, 104
Polyorchis, 155–7
polyp, 11–13, 162–4, 166
Polychaeta, 48, 55, 72, 87, 88, **107–14**, 122–4, 176–9, 182, 183
Polycladida, 18, 19, 22, 115
Polyphysia, 27, **44–6**, 72, 111, 180
Pomatias, 28, 41
Porifera, 163
potential energy, **15**, 61, 63, 66, 73, 154, 155, 158
powerstroke, 110, 112, 116, 117, 123, 124, 179
Precambrian period, 161
pressure, **7–10**, 32–4, 43, 51, 56, 67–9, **76–80**, 110, 117, 125, 133, 135, 154, 172, 182, 185
pressure pulse in, *Arenicola*, 49–52, **56–60**, 78–80
Ascaris, 117
Ascida, 14, 15
Chlamys, 152, 153
Ensis, 68, 69, 76–8
jellyfish, 156, 157
Loligo, 130, 132, **135**, 136, 139, 145, 149, 157
Lumbricus, 60, 62, 79
Metridium, 11, 12
Mya, 93
Octopus, 136–8
Peachia, 56
Petricola, 90–3
Priapulus, 55, 72
Sepia, 135, 136
Sipunculus, 55, 72
Venerupis, 90–2
Zirphaea, 100, 101
pressure transducer, 3, 9, 134, 153
Priapulus, 7, 54, 55, 71, 72, 78, 175
probing, 48, 51, 55, **64–71**, 76, 77, 81, 82
proboscis, 169–76
of *Arenicola*, **49–52**, 54, 55, 57–60, 70, 71, 74
of *Cerebratulus*, 72
of *Nephtys*, 112
of *Priapulus*, 7, 54, 55, 71, 72
of *Saccoglossus*, 70, 73
of *Sipunculus*, 7, 55, 72
of *Urosalpinx*, 89
propagated waves, 25–7, 41
propellor surface, 2, 18, 115
Prosobranchia, 88, 104, 168, 184
prostomial horns of *Polyphysia*, 72
protocoel, 176
Protostomia, 162, 170, 171, **176–9**, 186
Protozoa, 161–4
protractor muscle, 28, 63–7
Psammohydra, 167
pseudocoel, 118, 162, 166, 173, 174
pseudocoelomates, 162, 173, 174
pseudofaeces, 100, 101
pseudometamerism, 183, 186
Pseudostylachus, 88, 104
Pterobranchia, 171, 174, 177, 179, 186
Pteropoda, 117
pulmonata, 28, 168, 184

radial cleavage, 176
radial muscles, **143–50**, 155, 156
radial symmetry, 167, 168
radial thrust, 118, 181
radula, 89
razor shell, 66, 78, 81
recapitulation, 163
recording techniques, 9, 134, 144–7
recovery stroke, 123, 156
reproductive system, 174–6, 179
resistance of substrate, 70, 118, 119
respiration, 18, 135, 139, 144–6, 157, 168, 172, 183
resting condition, squid mantle, 147, 148
resting pressure, 11–13
retractor muscles, 11, **43–6**, 51, **63–70**, 72, 76–8, 90–3, 95, 110, 185
retractor strength, 69, 76
retrograde waves, **25–35**, 40, 41, 111, 122, 168, 179–84
Reynolds number, **19**, 115
Rhabdocoela, 18, 164
Rhynchodemus, 23, 24
rhynchocoel, 72, 76, 169, 170
rock boring, **89–101**, 104–6

rocking motion, 68, 93, 94
Rotifera, 173

Sabella, 113
Saccoglossus, 70, 73
salp, 130, 158
sand, 41, 52, 53, 56–8, 63, 67, 73, 74, 76, 81–5, 90, 112, 118, 159, 169
Saxicavacea, 89, 90
scallops, 129, **152–5**, 158
Scaphopoda, 73
schizocoelous development of coelom, 170, 172, 176
scraping device, 49–51
Scyphozoa, 157
sea anemone, **10–15**, 42, 44, 48, 56, 76, 78, 150
sea hare, 150–2
segments, **25–7**, 78–80, 179, 183, 185–7
segmentation, 124–8, 174–6
Sepia, 115, 130–3, 135–7, 143–9
septate worms, 8, 25–7, 60, 75, 175, 179
septum, **60–3**, 77–80, 128, 175, 180, 181
shell, 31, 48, **63–70**, 73, **87–102**, 105, 152, 161, 182, 183
shell muscle, 63–5
shell shape in burrowing, 69–70
shipworm, 102–4
siliceous spicules, 88
siphonoglyph, 11
siphonal retractor muscle, 93, 94, 100
siphons, 14, 15, 64–6, 81, 82, 85, 91–4, 97, 99–102
Sipunculida, 76, 87, 104
Sipunculus, 54, 55, 72, 78, 175
skeleton, axial, 1, 124–8, 187
exoskeleton, 1, 31, 74
fluid, 1, 10, 174, 187
skeletal flexure, 125–8
sliding friction, 38, 39
slip, 118, 119
slug, 35–8, 184
snail, 28, 89, 184
snakes, 117, 118, 122
soft substrates, 42–86
soil nematodes, 118
sole of foot, 30–41
Solemya, 68
spermatozoa, 115, 117, 122
spiral cleavage, 176, 182
spiral muscles, 30
sponge, 87, 88
squid, 3, 4, **130–7**, **139–45**, 147, 149, 154, 157, 158
stages of digging cycle, 66
static reaction, 19, 38, 39
stellate ganglion, 132, 147, 148
step length, 21, 22
stomodaeum, 180
Strombus, 40, 41
substrate, **42–106**, 111, 127, 167, 179
sucker, 21, 22, 180
surf clam, 80–6
swimming, 107–28, 150–7, 175, 187

tail, 25, 127, 168
technique, experimental, 9, 134
Tellina, 66, 70
Tellinacea, 63
tensile forces, 10, **76–8**, 92, 126–8, 135–8, 140
Terebella, 177
Teredo, 95, 102–5
terminal anchor, 43, 44, 48, 54–7, 60, 64–6, 70–6, 160, 170
test, 14, 15, 158
Thais, 88
thixotropic properties of substrate, 51
thrust, 50, 111, 114, 120, 129, 150, 153, 168, 181, 182
torsional force, 127, 128, 175
transducers, 3, 39, 70, 75, 77, 79
transverse muscle, **29–36**, 43–4, **63–9**, 72
trellis-like fibres, 17, 66, 117, 143
Trematoda, 164, 166
Tricladida, 18–24
triclads, terrestrial, 22–4
Tridacnidae, 89
Tridacna, 105
triploblastic animals, 16, 163, 164
Tritonidae, 89
trochophore larva, 176, 177, 182, 186
trunk coelom, 8, 49, 50, **56–8**, 71, 78, 173–5, 177, 178
tube, 113
Tunicata, 187
tunnel, 103, 104
Turbellaria, **16–24**, 28, 88, 104, 115, 117, 164–9, 183, 186
turbulent flow, 122
turgor, 21, 180
undulatory propulsion, 44, 70, **113–28**, 175–6
unicellular animals, 162
Urochordata, 171, 186, 187
Urosalpinx, 89

valvular mechanism, 130–2, 158
velar aperture, 156
velocity, 19, 115, 118, 122, 123, 130, 132, **140–2**, 153, 154, 157
velum, 152, 153, 155–7, 185
Veneracea, 89–91
Veneridae, 48
Venerupis, 91, 92
ventricle, 184, 185
Venus, 66, 184
Vertebrata, 1, 124–8, 171
vibration, 81
viscous forces, 16, 18, 19, 115, 118, 119
Volvox, 163

wash zone, 80, 83–6
water jet, **64–9**, **132–7**, 150–4, 158
waves, locomotory, **22–41**, 111–27
wood boring animals, 95, 102–4
work, 4
worm, **4–6, 16–22**, 25, 31, **43–52**, 76, 87, 88, 107–14, 117–19, 122–4, 142, 160, 174, 183, 186

Xylophaga, 105

yaw, 122

Zirphaea 90, **95–101**, 105